DIGITAL GROUND

Malcolm McCullough

DIGITAL
GROUND

Malcolm McCullough

Architecture, Pervasive Computing, and Environmental Knowing

The MIT Press   Cambridge, Massachusetts   London, England

First MIT Press paperback edition, 2005

© 2004 Massachusetts Institute of Technology

MIT Press books may be purchased at special quantity discounts for business or sales promotional use. For information, please email <special_sales@mitpress.mit.edu> or write to Special Sales Department, The MIT Press, 55 Hayward Street, Cambridge, MA 02142.

This book was set in Sabon and Kievit by The MIT Press and was printed and bound in the United States of America.

Research for this book was supported in part by the Graham Foundation for the Advancement of the Arts

Library of Congress Cataloging-in-Publication Data

McCullough, Malcolm.
  Digital ground : architecture, pervasive computing, and
  environmental knowing /
  Malcolm McCullough.
    p. cm.
  Includes bibliographical references and index.
  ISBN 0-262-13435-7 (hc. : alk. paper), 0-262-63327-2 (pb.)
  1. Computer architecture. 2. Human–computer interaction.
  3. System theory I. Title QA76.9.A73M45 2004
004'.35—dc21   2003056102

10 9 8 7 6 5 4 3 2

For my parents, who set me on maps.

Contents

Preface

This book has come from a change of outlook. It does not presume to be an impetus of such change for the reader, and is only an invitation to share in the author's inquiry. In some larger picture, many of us have been rethinking the relationship of environment and technology. Although this transformation seems most profound in biological fields such as ecology or public health, it also affects the disciplines involved in building social infrastructure. My own background is there, among people who work on architecture and computing. A decade ago, some people expected those fields to converge into something called cyberspace. Today, hardly anyone seems content with that notion. For me, and not me alone, part of the change has been a turn from the fast and far-reaching to the close and slow. I sit on stones in the sun more often, and hunch over screens in dark rooms less. Nevertheless the net still reaches me, wherever I am. That too has changed.

Few of us topple our viewpoints voluntarily, without a catalyst. For me, and for this book, that catalyst has been pervasive computing. This expression represents a paradigm shift from building virtual worlds toward embedding information technology into the ambient social complexities of the physical world. This shift deserves plentiful explanation and considerable skepticism, for while it has advantages in making technology more intuitive by means of embodiment, it also obviously has disadvantages in unwanted annoyances and surveillance. Frankly, to devote years of research and writing to so questionable a topic has often felt like folly. But this troublesome topic has merely been my catalyst, and not my pursuit. Although this book may first appear to be about information technology, it is ultimately a defense of architecture.

My claims about architecture are indirect because the design challenge of pervasive computing is more directly a question of interaction design. This growing field studies how people deal with technology—and how people deal with each other, through technology. As a consequence of pervasive computing, interaction design is poised to become one of the main liberal arts of the twenty-first century. I wrote this book because I ran into many people who believe that. If you share this belief, or if you just wonder what interaction design is in the first place, you may find some substance here in this book.

Although design writing often takes the form of advocacy or manifesto, that was not my intent here. Like the web logs that became fashionable during the time of this writing, my goal was only to sound some depths, connect some increasingly related disciplines, and provide some paths through a complex field. A book provides a more enduring form than a website, of course, and it invites a more thoughtful response. So despite the many doubts you may well have about its topic, I sincerely hope that reading this book can be enjoyable and worthwhile.

Acknowledgements

I must acknowledge the patience of anyone who has encountered me at work on this project, which of all the things I have ever completed is the one that least wanted to be. Although I cannot identify the true source of this difficulty, I must include how the explosive rise of the Internet has radically transformed the life of the solitary scholar. Although the university still expects isolated brainwork of its professors, the Internet makes it nearly impossible for work on any but the most narrow primary data-gathering to seem original, comprehensive, or complete. Those of us who were "classically" educated in the arts and the sciences may be especially vulnerable to the accelerating flow of knowledge. When the supply of secondary source material (the use of which distinguishes scholarship from other forms of research) becomes infinite, must its interpretation be reduced to a mere web log? Or may there be a new role, and taste, for the venerable act of concentrating a set of connected ideas into the relatively permanent form of a book?

Thus I am thus doubly thankful to those who helped me believe that this book was a legitimate endeavor. First among those was Richard Buchanan, who as head of the design school at Carnegie Mellon opened the most insights into that remarkable institution for me, and to the nature of argumentation as well. I am also particularly grateful to John Thackara (and his associates at Doors of Perception) for providing me with opportunities in Europe at time when America was culturally decelerating, and for voicing "the design challenge of pervasive computing" so well that I must echo his introductions as if they are now common wisdom. Third, I give heartfelt thanks to the Graham Foundation for Advanced Studies in the Fine Arts, whose grant in 1999 was pivotal in the formation of this project.

I am grateful to my colleagues at the University of Michigan for giving me the more stable home that proved necessary for completing this work. Almost every page of the result has been written or rewritten there—most of them, it must be said since the influence of context on activity is my topic, in the main hall of its magnificent law library.

Various artists and institutions furnished images, as credited in the text. Katherine Almeida oversaw two rounds of editing, before which the manuscript was unready. Yasuyo Iguchi has contributed a delightful design.

Several generous readers responded to murky drafts written at Carnegie Mellon in 2000 and Michigan in 2002. Thanks to Janet Abrams, Richard Buchanan, Paul Dourish, Debra Edelstein, Tom Erickson, Bridget Johnson, Andrea Klein, Edward Lee, Bruce Lindsey, Margaret McCormack, William Mitchell, and Andrea Moed. Thanks, and apologies, to any other readers whom I may have forgotten to mention.

My wife, Kit, was unremittingly sane, gave me the space that a writer needs, and shared my interest in exploring and examining the built world. My son, Callum, came into the world amid this project; and although he delayed it, he introduced so much lightness that eventually the work took better form.

Who could want computers everywhere? The saturation of the world with sensors and microchips should be a major story, and an active concern for all designers. Yet for several reasons, it is not.[1] Ever since the boom and bust of the Internet, many people feel that they have heard enough about computers. Since the events of 9/11, people fear that new roles for information technology are mostly about surveillance. Even without its political ramifications, the accumulation of "smart" devices in our lives increasingly seems misguided. Who wants more gadgets to own and manage—even ones that do not watch us? Who programs all this apparatus, and who among us can accept just how things have been programmed? Do smart machines generally force humans into stupid activities? These and other forms of information pollution have generally taken the shine off information technology. Meanwhile, the technofuturist imagination seems to have abandoned computing for biology. This is the century of the gene.

To turn our backs on computing would be foolish, however. To neglect further prospects in ambient, saturating, information technology will not make them go away. We would be wiser to accept them as a design challenge, to emphasize their more wholesome prospects (which are less likely to develop by default), and to connect them with what we value about the built world.

This is fundamentally a matter of embodiment. Digital networks are no longer separate from architecture. Unlike cyberspace, which was conceived as a tabula rasa, pervasive computing has to be inscribed into the social and environmental complexity of the existing physical environment. Situated technology may help us manage the protocols, flows, ecologies, and systems that form the basis of valued places; or it may add a layer of distrust, information glut, and experiential uniformity to them.

Unless this new field is to belong solely to technocrats, or tyrants, it demands richer cultural foundations. Here in the early 2000s, this design challenge appears to be in play. Interface design has become interaction design, and interaction design has come into alliance with architecture.

Human life is interactive life, in which architecture has long set the stage. The city remains the best arrangement for realizing that human nature. But today information technology allows people to interact remotely, asynchronously, and indirectly. Digital systems that are carried, worn, and embedded into physical situations can fundamentally alter how people interact. Psychologists, ethnographers, architects, and cultural geographers barely understand the consequences of all this mediation in terms of their respective disciplines, much less the implications for any new synthesis in design.[2] Where design has occurred at all, it has been without these cultural considerations. Software engineers have pursued the accumulation rather than integration of technical features. Interface designers have emphasized first-time usability at the expense of more satisfying long-term practices.

Notions of what a computer is have not kept pace with realities of how digital systems are applied. As ambient, social, and local provisions for everyday life, those realities have become part of architecture. Whereas previous paradigms of cyberspace threatened to dematerialize architecture, pervasive computing invites a defense of architecture. In sum, my essential claim is that interaction design must now serve our basic human need for getting into place.

In this book I have attempted to weave several threads of consensual wisdom into a common basis for architecturally situated interaction design. This foundation incorporates the ideas of philosophers on embodiment, psychologists on persistent structures, architects on scale and type, engineers on embedded systems, cultural geographers on infrastructures, and environmental economists on the value of places. Although the word *ground* may represent this broad basis as well, the book's title more specifically represents how information technology must be moved from the center of our focal attention into the periphery; and conversely, how certain contexts become responsive through the addition of technology.

In the book's three sections, I examine embodiment in context, technological issues in context-aware computing, and cultural frameworks for the value of context-oriented technology design. I have assembled arguments from architecture, psychology, software engineering, and geography to build a theory of place for interaction design.

Part I examines predispositions. What do we expect technology to be used for, what do we expect using it to be like, and how does any of that use enhance what we expect of everyday life? Many of these predispositions arise from our physical location, our embodiment, and our architectural settings. Persistent structures remain essential to how people understand and use the world. This is relevant as a basis for the current information technology paradigm shifts beyond cyberspace, and as a foundation for new forms of interaction design.

Part II turns from ends to means. Chapters 4–6 lay out the fundamental components of pervasive computing technology in terms of the physical gear (hardware), the symbolic models (software), and the patterns of usage (applications).

Part III explores the value of context-centered design. Chapter 7 presents interactivity as a cultural and not only technical challenge. Chapter 8 attempts to establish a worthwhile concept of place, despite that word's usually sentimental connotations. Chapter 9, in an apparent departure but essential turn, justifies place-centered design in terms of natural capitalism and psychological economics. All of these chapters emphasize the relevance of architecture and point to the increasing significance of interaction design.

This book should be of interest to people working in a variety of fields. They include the following.

Practicing digital designers. People who invent and integrate interactive systems now make a case for design. They need to understand recent developments and next steps in solving current problems. They also need to fill in background knowledge. They are strongest on day-to-day issues, and expect to learn more about how recent developments increase the practical role of design. They understand design advocacy quite well, but must be persuaded that the situation is changing with regard to the economic reductionism that they always face.

Researchers and developers in contextual technology. People who deliver reproducible results with empirical rigor need to understand the legitimacy of design. They know they can no longer ignore it. How is design measurable and accountable? How do interaction designers establish a different outlook on what is legitimate knowledge? With regard to such topics, these people are looking for next steps on practical problem of systems integration. They bring much more knowledge of specific technologies, but may wish to see a bigger picture. They may not recognize meta-issues of design as proposition, however. They need to be convinced this problem is theirs too. They need examples as problem solutions. They want to know how to view their own issues as design problems.

Architects and urbanists. Architects, and those in related disciplines of the physical environment, need to become aware of the challenges and opportunities raised by ubiquitous computing. They need to understand where technology is going, and what it has got to do with architecture. They may have little knowledge of the new field of interaction design. They may wish to know its principles, at least, with a special interest in new operations and challenges it brings to the built environment. They have been conditioned to an outlook of continuing marginalization, however. Many architects have conceptual difficulty with the digital world, and they may not anticipate ubiquitous computing.

General readers. Many general readers maintain an interest in the history, criticism, and philosophy of technological change. They tire of futurism, however, and wish to think about technology in terms of social and environmental concerns. They see broad connections, and think that a good read should involve those. Their knowledge of history varies. They have no particular knowledge of design process. Their general interest lies in how times are changing, especially with respect to technology and environment. They will want to see potential for solutions, but not within immediate developments so much as a broader historical sweep.

When everything else has gone from my brain—the President's name, the state capitals, neighborhoods where I lived, and then my own name and what it was on earth I sought, and then at length the faces of my friends, and finally the faces of my family—when all has dissolved, what will be left, I believe, is topology: the dreaming memory of the land as it lay this way and that.
—Annie Dillard, opening passage to *An American Childhood*

I Expectations

1   Interactive Futures

How do you deal with yet another device? How does technology mediate your dealings with other people? When are such mediations welcome, and when are they just annoying? How do you feel about things that think, and spaces that sense? You don't have to distrust technology to want it kept in its place.

The new field of interaction design explores these concerns. The more that interactive technology mediates everyday experience, the more it becomes subject matter for design. Like the electric light that you are probably using to read this book, the most significant technologies tend to disappear into daily life. Some work without our knowing about them, and some warrant our occasional monitoring. Some require tedious operation, and others invite more rewarding participation, as in games, sports, or crafts. These distinctions are degrees of interactivity.

The need for interaction design has become especially acute with respect to computers, the first truly interactive technology. No longer just a tool for producing documents, networked computing has long since become a social medium. As interactivity pioneer Brenda Laurel declared in the early 1990s, "the real significance of computing has become its capacity to let us take part in shared representations of action."[1] These representations can be of organizations, activities, problems, work practices, communities of interest, and not just predictable numerical models. Some of these representations are coded explicitly, but at least as many remain implicit in the contexts and configurations of technology usage. Representations of work and play now become, in effect, the software of places. These need more intentional design.

Software engineers think they know what they mean by design, and so do architects. When information technology becomes a part of the social infrastructure, it demands design consideration from a broad range of disciplines. Social, psychological, aesthetic, and functional factors all must play a role in the design. Appropriateness surpasses performance as the key to technological success. Appropriateness is almost always a matter of context. We understand our better contexts as places, and we understand better design for places as architecture.

Like architecture before it, and increasingly as a part of architecture, interaction design now becomes a critical liberal art. However, to discuss such propositions is to get ahead of the story.

Ubiquity

At the theater one night, you might find yourself wanting to jam all the mobile phones in the house. Sliding through the EZ Pass lane on the way home afterward, you might notice how the idea of fixed devices interacting with mobile devices is not so unusual (figure 1.1).[2] Walking into your house preheated by its programmable thermostat, you might realize that just as much computation is built into your surroundings as is carried about in your bag.

Just as text long since escaped medieval monasteries and can now be found not only in portable books, but also on stickers, shirts, street signs, and all over product packaging, similarly computers have long since escaped the glassed-in laboratory and the beige office cubicle.[3]

We see them everywhere and sometimes they see us. Far more microchips go into objects we hardly think of as computers than into

1.1 Offloading information onto context. The mobile device meets the fixed and embedded device. (*Courtesy of Analia Cervini, Interaction Design Institute Ivrea.*)

boxes used through a keyboard, mouse, and screen. Today less than a quarter of the chips produced by Intel, the largest manufacturer, are put into desktop or laptop computer motherboards.[4] The rest are embedded into things that you carry about, drive, or wear; or they are embedded into physical locations. They drive personal gadgets, information appliances, smart tags, responsive rooms, environmental monitors, and location-based services.

Since about 1994, microprocessors have outnumbered humans on this Earth. As of 2002, for each person in the United States, there existed a microelectromechanical system (MEMS) chip, which is an essential component in physical-digital interfaces. Technology visionary Mark Weiser defined ubiquitous computing as "hundreds of computers per person." Also known as ambient, physical, embedded, environmental, or pervasive computing, ubiquity has succeeded cyberspace as Silicon Valley's party line on the technological future (figure 1.2). When the Association for Computing Machinery (ACM), the world's largest membership organization of information technology researchers, launched a general-readership publication named Ubiquity, and called its plenary conference "After Cyberspace," the paradigm-shift had become more or less official.[5]

Many of these terms have become overexposed. The word *interaction* has lately been applied to just about any relationship between people or things, as though shapes interact in a Picasso painting. More properly, the word implies deliberation over the exchange of messages.[6] Thus you don't interact with a book, you just read it. But using electronic communication, you can interact with other people who are not physically present, or who take part in the interaction at some other time. Thus through digital media, we interact indirectly.

Similarly the word *ubiquity* was seldom heard until recent years, but now is applied to all manner of globalizing technology. Within the continual noise of technology hype, and like the word cyberspace before it, ubiquity has quickly come to mean just about anything having to do with universal connectivity. To people still catching up with the Internet, ubiquitous computing seems to mean wiring up every last seat in their workplace, or wirelessly browsing the Internet from any location on Earth.

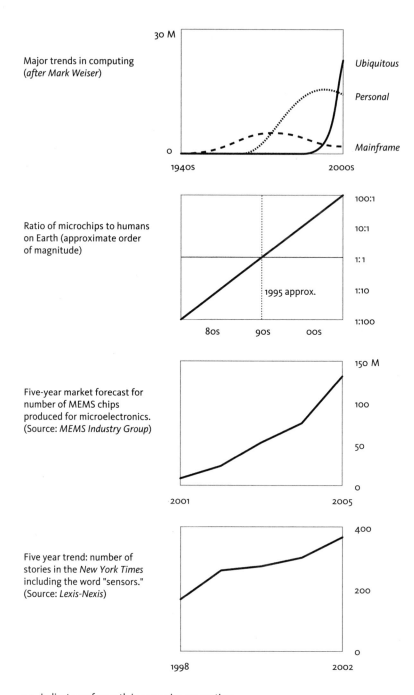

Major trends in computing
(*after Mark Weiser*)

30 M

Ubiquitous

Personal

Mainframe

0

1940s
2000s

Ratio of microchips to humans
on Earth (approximate order
of magnitude)

100:1

10:1

1:1

1:10

1:100

1995 approx.

80s
90s
00s

Five-year market forecast for
number of MEMS chips
produced for microelectronics.
(Source: *MEMS Industry Group*)

150 M

100

50

0

2001
2005

Five year trend: number of
stories in the *New York Times*
including the word "sensors."
(Source: *Lexis-Nexis*)

400

200

0

1998
2002

1.2 Indicators of growth in pervasive computing

To emphasize the invisibility of chips in everyday things, the word *pervasive* has become more usual. According to a characterization from the year 2000 by the National Institute for Standards and Technology pervasive computing is "(1) numerous, casually accessible, often invisible computing devices, (2) frequently mobile or embedded in the environment, (3) connected to an increasingly ubiquitous network structure."[7] Intel announced the technological future at the turn of the millennium:

Computing, not computers will characterize the next era of the computer age. The critical focus in the very near future will be on ubiquitous access to pervasive and largely invisible computing resources. A continuum of information processing devices ranging from microscopic embedded devices to giant server farms will be woven together with a communication fabric that integrates all of today's networks with networks of the future. Adaptive software will be self-organizing, self-configuring, robust and renewable. At every level and in every conceivable environment, computing will be fully integrated with our daily lives.[8]

Project Oxygen at the Massachusetts Institute of Technology presented a similar picture (figure 1.3):

In the future, computation will be human-centered. It will be freely available everywhere, like batteries and power sockets, or oxygen in the air we breathe. It will enter the human world, handling our goals and needs and helping us to do more while doing less. We will not need to carry our own devices around with us. Instead, configurable generic devices, either handheld or embedded in the environment, will bring computation to us, whenever we need it and wherever we might be. As we interact with these "anonymous" devices, they will adopt our information personalities. They will respect our desires for privacy and security. We won't have to type, click, or learn new computer jargon. Instead, we'll communicate naturally, using speech and gestures that describe our intent ("send this to Hari" or "print that picture on

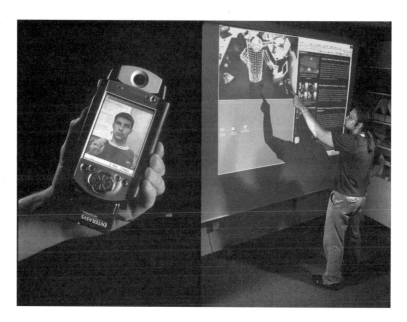

1.3 Recognizing faces and poses in Project Oxygen, MIT's initiative about ambient and ubiquitous computing. (*Courtesy of Project Oxygen.*)

the nearest color printer"), and leave it to the computer to carry out our will.[9]

Business Week, in its "21 Ideas for the 21st Century," said:

> In the next century, planet earth will don an electronic skin. It will use the Internet as a scaffold to support and transmit sensations. This skin is already being stitched together. It consists of millions of embedded electronic measuring devices: thermostats, pressure gauges, pollution detectors, cameras, microphones, glucose sensors, EKGs, electroencephalographs. These will probe and monitor cities and endangered species, the atmosphere, our ships, highways and fleets of trucks, our conversations, our bodies - even our dreams.[10]

So much future tense just annoys many of us. Consider why.

You may have thought that the future of computing was all about virtual worlds, intelligent agents, and cyberspace. If you travel among technical circles, you may have been hearing much lately about nanotechnology and breaking down the hardware-software barrier. Or perhaps you were just dreaming of the paperless office. These are examples of technological futures. Expectations about the role of technology seem especially important to designers.

Recently we have witnessed a paradigm shift from cyberspace to pervasive computing. Instead of pulling us through the looking glass into some sterile, luminous world, digital technology now pours out beyond the screen, into our messy places, under our laws of physics; it is built into our rooms, embedded in our props and devices—everywhere.

This may not impress anyone who conflates either of these notions with the technologies of the Internet. Lumping these ideas together was common enough amid the economic frenzy of the 1990s, when these technologies were all new. Indeed if cyberspace were the Internet itself, hardly anybody would be referring to it in the past tense. What has passed is expectation for a coherent there, there; the chaotic reality of the Internet lives on. When the pundits of Silicon Valley do use the past tense, they are referring to a unifying futurist paradigm, and not to what remain very viable applications of spatial data visualization, networked organizational change, and online community building.

At least in the popular imagination, cyberspace consisted of the notion that the Internet was a coherent place apart that you could immersively inhabit. This "consensual hallucination," as it was so often called, was more than a metaphor, and at times seemed more like a societally enacted myth. That is an instance of a technofuture. When by late 1996 no less august an institution than the New York Times not only discussed but promoted it, cyberspace had become a household word. Apparently everyone believed there was a there, on the other side of the looking glass.[11] In a full-page ad for itself, the Times proclaimed a new civics. "It's part newspaper, part gathering place. . . . In a world as complex as the Web, it's reassuring to know there is, in fact, a town square."[12]

It might as well have been puppet theater. You entered the net through the looking glass of the computer screen, or goggles, and there, at the edge of the aided senses, you saw ephemeral projections of things from higher, more abstract realms. You imagined "visiting" sites when in fact your browser software downloaded packets of data to wherever you were sitting.

This disembodiment had its limits. Suspension of disbelief didn't make it to the inner ear. To most of us who experienced immersive virtual reality at any length, the result was "simulation sickness": nausea induced by the disconnect between reverse-engineered visual space and our bodily kinesthetic orientation systems.

Meanwhile at a more practical level, the management consultancies have generally established how the knowledge that computer systems aim to represent resides in communities, organizations, and physical arrangements of props and devices. Protocols, so essential to the social role of the net, remained a function of embodiment. And usability, that foremost goal of interface design, proved to be less a question of immersion than of embodied activity in habitual context.

When a term spreads through a culture quickly, it often represents a passing wave of seeing the unexplainable world in some particular way. Then the metaphor wears off. The need to explain new technologies in terms of older realities generally tends to diminish. Thus automobiles eventually ceased to be horseless carriages. "Ban cyberspace," ran an *Economist* headline in June 1997, "The word, not the thing itself, whatever it may be. And dump the rest of the Internet's lame metaphors too."[13]

The cutting edge dulls on everyday life. Often the technologies on which new expectations are based blend into the fabric of everyday existence. Like the telephone before it, for instance, the Internet has begun to fade into banal, unlovely normalcy. Other technologies are rejected for errors in principle. Much as bloodletting turned out to be inaccurate in medicine, so virtual reality left out some important details—such as the fact that we orient spatially not just with our eyes, but also with our body.[14] Then too, other technologies are rendered obsolete by unforeseen alternatives, as freight trains were by interstate trucking.

Ubiquitous computing, in its universalist version has overlooked the value of context. Humanity has had thousands of years to build languages, conventions, and architectures of physical places. Wave upon wave of technology has transformed those cultural elements, but seldom done away with them. Context appears to have unintended consequences for information technology.

Meanwhile, the disembodied quality of global digital information flows has become a source of fear. Like some medieval map with monsters rampant in its margins, popular notions of cyberspace involve some dark fear of terra incognita.[15] Especially to those left on the wrong side of the digital divide, which was effectively represented by the monitor screen, the innate response was distrust. Cyberspace was dark, it was vast, and it was full of tricksters.[16]

As a form of urbanism, cyberspace was perhaps also some last version of what is sometimes called the project of transparency. At the risk of oversimplifying the academic significance of this word, in this context transparency describes how the idealized modern city sought to overcome the squalor and Victorian ponderousness of the industrial city with light and motion. Modernity espoused the belief that humanity must remake the world according to its own rational abstractions. It sometimes quite literally paved over anything that detracted from the predictability of its methods. As the modernist city was primarily a response to fears of disease, disorder, and impenetrable density, so fear of dark space led city planners to bulldoze entire neighborhoods in the name of the open plan.[17]

A similar ambition may explain cyberspace imagery, at least in the seminal Gibsonian version: "panoptic windows onto unified, glowing urbanistic infrastructures of flow.... A graphic representation of data abstracted from the banks of every computer in the human system. Unthinkable complexity. Lines of light ranged in the nonspace of the mind, clusters and constellations of data. Like city lights, receding."[18]

Much as by the 1960s, dissenting writers like Jane Jacobs and Aldo Rossi were explaining how the modernist city was all wrong, now current work in situated, embodied interaction design questions the aims of universalizing, disembodied, cyberspace. We now see harm in the belief that computer usage must be the same everywhere, and that it will transport us to some fantastical otherworld. Instead, much as Jane Jacobs found living service ecologies in the apparent chaos of premodern neighborhoods, interaction designers now turn to the patterns of the living world as something other than a clean slate, and something to be understood, not overcome.

To illustrate this fundamental shift away from the limits of world making, consider a more detailed story. This concerns that most emblematic instance of urban technofuturism, The World of Tomorrow, the 1939 New York World's Fair.

The fair's identity remains highly recognizable to many of us. Its identity still receives a lot of attention, from revisionist historians to television cartoons.[19] There has been a lot of swooping architecture lately that recalls its buildings. The fair's memorable combination of unornamented curving surfaces, transparent cockpits, and supremely confident newsreel voiceovers has become an emblem of futurism. Its domestic science fiction has since been topped in outrageous imagination by the 1960s British mod architecture magazine *Archigram* and 1990s cyberpunk novels such as *Snow Crash*, for example, but the fair's popular impact may remain without equal.

As introduced in its official guidebook, The World of Tomorrow was a high-water mark of belief in worldmaking. "The Fair you are now enjoying is the result of the happy combination of the dreams, the experience, and the courage of many men and women," it declared. "The true poets of the 20th century are the designers, the architects, and the engineers who glimpse some inner vision, create some beautiful figment of the imagination, and then translate it into valid actuality for the world to enjoy.[20]

"The story we have to tell," the urbanist and historian Lewis Mumford remarked at a fair strategists' dinner, "and which will bring people from all over the world to New York, not merely from the United States, is the story of this planned environment, this planned industry, this planned civilization."[21]

Most of all, the fair illustrated that modern space was all about freedom of movement. "In order to grasp the true nature of space the observer must project himself through it."[22] As promulgated through much of the twentieth century by Sigfried Giedion's canonical text-

1.4 Modern space and technofuturism: Futurama, the General Motors pavilion at the 1939 New York world's fair. (*Courtesy of the Norman Bel Geddes collection, Harry Ransom Center for the Arts, University of Texas.*)

book *Space, Time, and Architecture*, which was published two years after the fair, the fundamental act of design became the shaping not of buildings, but of space. Flow itself now became the fundamental concern of the architect, not only for people and their vehicles, but for conceptual space. As in cubism, and in a rather misguided interpretation of Einstein, space became conceived in relation to a moving point of reference. It is no accident that Futurama, the fair's centerpiece, was created by a transportation company (figure 1.4). No wonder that in the decades to follow the traffic engineers were effectively given full power over the form of the city—and no wonder that the most dumbed-down metaphors for the Internet are automotive.

In sum, as modernity remade the world, at almost every instance it preferred motion to rest, the open instead of the concealed, and control rather than complexity. Just how far off the mark all this overconfidence really fell was perhaps best observed by the master essayist E. B. White commenting for the New Yorker on Futurama.

The countryside unfolds before you in $5-million micro-loveli-
ness, conceived in motion and executed by Norman Bel Geddes.
The voice is of utmost respect, of complete religious faith in the
eternal benefaction of faster travel. The highways unroll in rib-
bons of perfection through the fertile and rejuvenated America of
1960—a vision of the day to come, the unobstructed left turn, the
vanished grade crossing, the town which beckons but does not
impede, the millennium of passionless motion. When night falls in
the General Motors exhibit and you lean back in the cushioned
chair (yourself in motion and the world so still) and hear (from
the depths of the chair) the soft electric assurance of a better life—
the life which rests on wheels alone—there is a strong, sweet poi-
son which infects the blood. I didn't want to wake up. I liked
1960 in purple light, going a hundred miles an hour around
impossible turns ever onward toward the certified cities of the
flawless future. It wasn't until I passed an apple orchard and saw
the trees, each blooming under its own canopy of glass, that I perceived that even the General Motors dream, as dreams often do,
left some questions unanswered about the future. The apple tree
of tomorrow, abloom under its inviolate hood, makes you stop
and wonder. How will the little boy climb it? Where will the little
bird build its nest?[23]

Accursed Computing

With the lessons of cyberspace and Futurama in mind, we turn to pervasive computing. The saturation of the world with sensors and
microchips should become a major story, and an active concern for all
designers, but so far it has not. Because digital technology was so over-
sold and overbuilt during the recent Internet boom, people no longer
want to hear about computing.[24] Because genetics have taken over as
the next big thing, people no longer look to computing for their deepest
cultural challenges. Compared with the possible consequences of decod-
ing the human genome, desktop computing seems clumsy and quaint.

Such neglect is unwise, for computing is hardly going away. No
longer the province of overenthusiastic young entrepreneurs, digital

technology has been left to governments and corporations with far less felicitious goals.[25]

Surveillance, for instance, has become an unfortunate fact of life. The loss of privacy has become a central theme in cultural studies of information technology.[26] Much as smog is objectionable but does not make us surrender our cars, surveillance is a bad side effect of information technology but it is not intrusive enough to make us give it up. Now as inexpensive cameras and sensors show up in more places, that side effect worsens. Our devices are watching us (figure 1.5). This is the foremost purpose and the most usual objection to pervasive computing.

One usual response to surveillance has been to revive Orwellian fears of some unblinking, totalitarian Big Brother. Although America, for one, has a deep concern in the erosion of civil liberties, the twentieth-century version of the panopticon may be outdated now.

First of all, omniscience is elusive. As anyone who has ever tried to resolve a simple billing dispute knows, even the telephone company lacks enough internal coordination to make sense of its data about you. And as anyone who has ever dealt with a state-level bureaucracy knows, the odds of omnicompetance remain low. Generally as information becomes more and more abundant, clear views through it become less and less possible.

Furthermore, there are a lot more parties doing the looking. Instead of Big Brother, this is more like ten thousand little brothers. For example, one order for radiofrequency identification tags that made the news in late 2002 involved half a million units. Besides having people nervous about privacy, pervasive computing raises concerns about the proliferation of autonomous annoyances. Does anyone want chirpy little advisors (such as the animated paperclip in Microsoft Word) to escape beyond the desktop and hit the streets? Instead of "Hi! You appear to be writing a letter!" you would have to put up with "Hi! You appear to be walking past our shop!" Nevertheless, even without speculation, we can observe plenty of annoyance in the form of petty information pollution.[27] It is muzak spewing out of gas pump handles. It is messages waiting in more devices than one cares to monitor. It is safety labels that pack text files.

1.5   One response to surveillance: Bhutto performance art by Daniel Beaubois:
"In the event of amnesia the city will remember." (*Courtesy of Daniel Beaubois.*)

Expectations

It is robotic pets, home automation systems, and relentless entertainment. Its purveyors assume no more responsibility for information pollution than nineteenth-century industrialists did for dumping sludge in the river. The assumptions behind its cultural ambition and its availability are made at the source, not the destination. Proactive information feeds treat all quiet time and space as something that needs filling. Portable and embedded devices take these streams out from your computer screen and into the world, where they are more difficult to turn off.

And then these devices crash. If a chair held you only 99 percent of the time, you would hesitate to sit on it. If a face-recognition security system mistook an identity just a thousandth of the time, terrible legal and social difficulties would result. If your car had some exciting new interface, it might be more dangerous to drive. Would you care to begin the day by reading a message that your house's software was down?

Finally, even before reliability becomes an issue, the programmability of physical-world systems has been the prime objection to pervasive computing (figure 1.6). We have neither the time to program so many systems ourselves nor the willingness to accept how others might program them for us.

Doors that swing open for you are one thing, but word processors that rewrite sentences or capitalize words on your behalf are another. Inflexibly configured systems might be tolerable in aesthetic matters such as lighting a room, but they are intolerable in critical matters such as medical equipment. Who programs all this stuff, how much of it can you reprogram, and how much programming and reprogramming can you stand? Since most of us can write little or no computer code, have to memorize far too many instruction sequences and passwords already, and lack time to learn how to operate even one more device, who is going to adopt smart technologies? How unobtrusively, even naturally, can all this activity occur? Few of us want our experiences designed for us; yet just about every one of our experiences that is mediated by technology could be better designed. It is to address this paradox of programmability that the new discipline of interaction design has emerged.

Yes,	But...
Computers are everywhere	Who asked for this?
Anytime-anyplace!	Equals nowhere
Objects will be smart	And they will force us to do something stupid
Anything can be on the internet	Do you need e-mail in a toaster?
Anything can get an interface	Will they all flash 12:00, like VCRs?
We can invent the future	But don't damage what already exists
Microchips are cheap	Dealing with them is expensive
Buildings get nervous systems	Inhabitants get nervous
You can monitor your family	Does that build trust?
Stuff becomes programmable	I don't have time
Stuff becomes programmable	I don't like the way someone else does it
Systems anticipate needs	And they assume we need entertainment
Tags can carry instructions	Mind the step; eat your vegetables
Systems respond to you	Hi!! You appear to be writing a letter!!
Smart conveniences	What, the curtains?
People won't tolerate this	Look how they took up mobile phones
Who could love a computer	Did the farmer love his plow?
Big brother is watching	Through terabytes of data smog
It's all about surveillance	And cars are all about emissions? bad side effects
Computers crash	So do cars, but we still use them
The net boom is over	Computers are not going away; quite the contrary
I'm against technology	Except my dishwasher

1.6 Common objections to pervasive computing

Interaction designers study how people learn, operate, and assimilate technology, especially information technology. They also study how technological mediation influences what people are doing. Sociologists, psychologists, and management consultants address such concerns as well, but at a more general level. In comparison to those disciplines, interaction designers emphasize the particular mechanisms of product usability. Increasingly, they do so in terms of work practices, social organizations, and physical configurations—in a word, context.

The use of the term interaction design instead of interface represents a cultural advance in the field. Recent mission statements by firms, schools, and publications commonly acknowledge this.[28] Interaction designers claim to know at least partly what is wrong with information technology, and that overemphasis on technical features and interface mechanics has been a part of the problem. By turning attention to how technology accumulates locally to become an ambient and social medium, interaction design brings this work more closely into alignment with the concerns of architecture.

Because architects and designers of noncomputer systems may be unfamiliar with the history of this field whose evolution now leads toward them, a brief overview of this progression may be helpful. If the current stage of computing becoming pervasive constitutes a milestone, it is worth comparing that stage with two others: first, the growth of machine interface design; and second, the achievement of machine interactivity.

In what is often cited as a starting point in the industrial design of interfaces, Henry Dreyfuss, a proponent of the new field (and incidentally a chief designer of Futurama) observed: "If the point of contact between the product and the people becomes a point of friction, then the industrial designer has failed. If, on the other hand, the people are made safer, more comfortable, more eager to purchase, more efficient, or just plain happier, then the design has succeeded."[29]

In contrast to present interests in software usability and participation in information flows, industrial interface design was more often addressed to automation. The early twentieth century imagination expected advances in interfaces to eliminate participation wherever possible.[30] This is relevant to us because early developments in information technology assumed that legacy. Symbolic processors are not actually moving mechanisms for the transfer of powered motion, but to this day we still call them "machines."

Interactivity changed the role of technology, however. In our review, this is the second milestone. The ascent of human-computer interaction as a design discipline required a fundamental shift in expectations. What made the personal computer so radical was the notion that someone might look forward to using it.

More specifically, computers became the first technology to provide two-way engagement. Despite common misuse of the word, not everything that is operable is interactive. A film may stir deep reactions; a chisel might let a sculptor feel that work is flowing; a lathe may have several buttons and controls; and a telephone lets people interact remotely; yet none of these technologies is itself interactive. Only when technology makes deliberative and variable response to each in a series of exchanges is it at all interactive. Such exchange is like a conversation in how participants coordinate process as well as content by means of acknowledgments, corrective interruptions, and cues. Although some people too readily attribute thought to symbolic processing technology, nevertheless we rightly experience interaction.[31] A computer might even beat you at chess.

Computer-human interface (CHI) became the subject matter of design only when processing and memory become inexpensive enough that they could be used not only to accomplish storage and calculations, but also to make those processes more convenient to people. The familiar graphical user interface (GUI) represents the latter stage of development. It is of course what first made computing accessible to nonspecialists. The admission of psychological principles into the previously all hard-numbers field of computer science brought it to the mainstream. Twenty years later, and still measured in mechanical first-time usability, building better interfaces remains the goal of much of the CHI community. (Not surprisingly, this community sometimes approaches ubiquity as if that means putting those window-and-menu screens everywhere.)

As interactivity become more widespread, expectations for automation gave way. For example, up until the network computing boom of the 1990s, efforts at artificial intelligence sought to capture knowledge, build inference engines, and, ever in industrialist mindset, proceduralize competent work. Then the spread of networks made information technology into a catalyst of organizational change. Designers and managers then recognized how the kinds of expertise resident in communities were unlikely to be automated, but could be served by better information "environments."

The idea of context has been growing all along. The graphical user interface was a conceived as a context for processing symbols, for instance. Later, the information flow through an enterprise was a context in which new software had to be introduced appropriately. Next that flow moved out onto mobile devices. Those devices meet up in arbitrary locations; others are embedded into relatively permanent local configurations; and sensors and effectors are added to the built environments that house them.

What is at issue is participation. The pushbutton industrial machinery of 1939 and the virtual realities of 1989 both left the human subject just sitting. Well-being requires a better state of human activity. Much of the human sense of environment emerges from our activity in habitual contexts. All this becomes the subject matter of design.

In the words of designer Clement Mok, "The most basic function of an interactivity art is providing a cue for a specific action."[32] Today the context of the digital task has extended beyond the desktop to world of work, play, travel, and dwelling. To anyone with too much gear and too little time, the mere availability of technical capabilities hardly guarantees utilization.[33] Whether features are understood and applied depends on context in which they are encountered. At this point, "contextual design" of information technology has to address such practices in situ.

This is the latest milestone. The role of computing has changed. Information technology has become ambient social infrastructure. This allies it with architecture. No longer just made of objects, computing now consists of situations.

A Cultural Challenge

Rather than turning our backs on pervasive computing because surveillance is objectionable or the Internet boom is over, we should explore its cultural aspects. We should no more ignore this movement than the Internet or personal computing before it. Given our fears about privacy, autonomous annoyances, and rigidly preprogrammed activities, we should pay more, not less attention to this stage of technological development.

As you fuss to assimilate yet another bit of hardware or software into your daily routine, such grand ideas may seem awfully distant. Like the videocassette recorder flashing 12:00 in living rooms all over the world, just about every addition of gear to our lives comes with more technical detail than we are ready to absorb. Some of it is just unnecessary. What if your latest car came with additional pedals on the floor?

Today we can no longer assume that mechanical efficiency is the root of usability, that more features mean better technology, or that separately engineered devices will aggregate into anything like optimal wholes. The kinds of judgment necessary for establishing appropriateness in interaction design are at least as professional as artistic or scientific in character. We need to advance the science of the computer-human-interface into a culture of situated interaction design (figure 1.7). "We" is a lot of us: psychologists, architects, ethnographers, product designers, entertainers, management consultants, policy makers.

This challenge seems inseparable from establishing more general legitimacy for design. When the most conservative accountancies are declaring the value of design, and more creative strategists are understanding design in terms of the propositional thinking that occurs beyond the limits of predictive analysis, then design, writ large, is becoming more important. Under this broader conception of design, better technology is not just faster, prettier, or more usable, although those attributes are usually welcome. It must also be useful, and it should also be more appropriate. Thus it must be the product of cultural deliberation. If it is not, then it is likely to be objectionable, and perhaps costly.

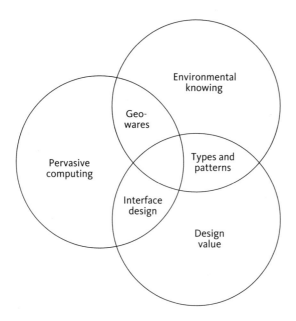

1.7 Intersecting domains

Fortunately, so far in the history of computing, the law of unintended consequences has tended more toward chaos, creativity, and
occasional delight, and less toward the sorts of command-and-control
anticipated in the industrial era. As in the first fun software of the
1980s, or the first online social lives of the 1990s, our present
decade's early delights in smart things and responsive spaces may
come from people not burdened by existing expectations about the
role of technology.

Expectations are critical. Expectation management dominates
technology implementations. What technology can do may not be so
important as what we want to do with it, and whether that is reasonable.

To modernity, technology was for world making: to overcome the
limits presented to us by our place in the physical world. Its goal has
been pure artifice.[34] With an unprecedented confidence in the accuracy of its methods, modernity has imposed its formulas on the world
until they have become the world. When it has worked, this approach

has relieved suffering and introduced convenience. When it failed, it was attempting to straighten rivers,[35] house people in high-rise filing cabinets, or plat political borders where no terrain features or language difference suggested them. Whether in government, corporations, or universities, decision makers have become so caught up in modernity's mechanistic beliefs that they reject most appeals to nature.[36]

What is missing in the World of Tomorrow, or its latter-day counterpart in cyberspace, or in the anytime-anyplace version of ubiquitous computing, is the world itself. *Homo faber* has an Achilles' heel; his artifice cuts him off from his nature. This is a fatal separation. In the oft-quoted words of the landscape architect Ian McHarg: "No species can survive in an environment solely of its own making."[37]

Now as environmental limits pronounce themselves more loudly, however, the last century's headstrong attitude toward world making must eventually give way. Under present global environmental circumstances, appeals to place can no longer be dismissed as romanticism.

As the discipline of interaction design continues to mature, it must be measured by increases in human, cultural, and natural capital. It must involve more kinds of observation and critique. As graduate programs sprout in universities, let their proponents find a way beyond business automation. If communication technologies affect imaginations, let there be an awakening of mental environmentalism. Since cultural productions are measured in appreciation, let interactivity inspire staff critics to write weekly columns in the local newspaper.

But let us avoid the future tense. Let us focus on habits rather than novelties, on people rather than machines, and on the richness of existing places rather than invention from thin air. What purpose do we expect pervasive information technology to serve? When, if ever, does it seem natural to use?

2 Embodied Predispositions

Place begins with embodiment. Body is place, and it shapes your perceptions. Embodiment is not just a state of being but an emergent quality of interactions.[1]

The discipline of interaction design has been built from foundations in our understanding of cognition. Increasingly, this work recognizes the importance of "cognitive background": the cumulative perceptions of enduring structures that fundamentally shape human abilities.

The discipline of architecture also reflects some deep knowledge of environmental perception. This, more than fashionable geometry, is what that older field may best contribute to the newer one. For interaction designers seeking to know more about context, space, and place, and conversely for architects wishing to understand the roots of interactivity, the principles of embodied predispositions provide increasingly common ground. For anyone wishing to understand the role of context, a detailed look at these foundations is worthwhile. Principles generally acknowledged by environmental psychologists, applied by architects (and occasionally dramatized by body artists), now become relevant to the design of information technology. Any review of these theoretical principles is necessarily dense and may appear too academic at times, but it provides a useful foundation for new technological developments in contextual awareness. It also provides one basis for the current shift from virtual world building to pervasive computing.

The exploration of embodied interactions reveals to us conditions otherwise often taken for granted, yet to study them is not to state the obvious. Many of these conditions are familiar to each of us, but difficult to predict or measure. Nevertheless we cannot dismiss them for lack of metrical proof. They may be essential to experience, yet we need not conflate them with questions of pure philosophy.

To be practical, this inquiry must emphasize everyday situations. For example, consider the corner office. Obviously, occupying this location is an expression of power, which comes with practical rewards of more light, views, and air than other offices have. It also functions as a site of exchange, for although information can be transmitted in the abstract, the exercise of status still demands a chance for the players to size each other up. Hence the better corner office provides a variety of locations in which two or three people can sit in relation to each other. Body language matters here. Someone might get hot under the collar. This demonstrates that even in us well-dressed mammals, visceral factors such as gesture, temperature, and smell still influence the establishment of rank.[2] This alone guarantees that although telecommunications have taken over information exchange, they may never replace face-to-face meetings for exchanging power and opportunity. Although the protocols used for human meetings are more subtle than circling and sniffing, this exercise of power still depends on physical factors. As evidence for how sensitive humans are about the spaces they use for negotiation, consider how diplomats haggle over the shape of council tables, or salesmen carefully size their closing rooms. Appropriately configured physical space tacitly allows subtle variations in interpersonal distance.

From one person's body language to a whole society's body politic, much else besides the exchange of power depends on embodiment in contexts. Lighthearted conviviality works best at close quarters over food. A personal dwelling involves an accumulation of tangible souvenirs. Societal memory uses physical landmarks, and this is what makes the city the repository of civilization. Social recreation uses public sites for the presentation of self, for which physical architecture sets the stage. A rich and deeply structured background of environmental patterns exists not only in the individual but also in the culture and the species. The exercise of these is the making of the world.

Body Image, Body Art

The body is your first and last site, your center, and your scale. As stated best by the cultural geographer Yi-Fu Tuan, "the body imposes a schema on space."[3] Up is most decidedly different from down; front is different from back; the world unfolds before us and recedes behind us.[4] We move forward. To confront a problem is the opposite of turning your back on it. Left is even different from right, despite that being our one essential axis of bodily symmetry. People turn right more easily, at least in America, and shopping malls sell more goods off the walls that are on customers' right side as they enter the stores.

Besides giving orientation, this bodily schema establishes range. The distribution of guests at a cocktail party demonstrates distance. Range can be aural—close enough to listen or far enough to be out of earshot—or a much more complex matter of social protocols based on personal familiarity. Bodily range also incites action. Something within reach of your strides suggests that you might move closer. Things within your reach take on greater significance and are perceived more vividly and actively than things far away. Things within your grasp invite use.[5]

Along with range, the body gives scale. Whether something is relatively larger or smaller than you are affects how you react to it. The same picture reads differently at poster and at postage-stamp sizes.[6] Objects and spaces near our own scale are more comforting than abstract ideas and measurements at radically different scales.

Much of our bodily stress comes from encounters with the vast or the tiny. The body gives scale, shape, and orientation to our picture of ourselves in the world. Astrophysics and microbiology distress some of us because they demonstrate that human scale is not the measure of all things. In contrast to all that is neutral, infinitely extensible, isotropic, and empty about rational "objective" space, embodiment is highly subjective. Extremes of scale conflict with the image that the body is a center.

Body image is actively constructed.[7] Social games play at presentation of this self in different environments. Sports and dances cultivate the abilities of centering. In the practice of Tai Chi, for instance, one works to move the center of one's body image (which has often crept upward toward the head) back down into alignment with one's kinesthetic center. Such interrelation of sensation, motion, posture, and expression occurs unconsciously throughout life's processes, and from these relations each of us builds a keen sense of haptic orientation. (The word *haptic* describes the active, probing aspect of the sense of touch.) Haptic orientation often precedes the formation of visual mental models, and is important to the study of predispositions.[8] Such precedence is at work when you jump out of your seat at a movie, for example.

The recent surge in body art tells us that embodiment has been in play in both street and gallery culture.[9] Visceral art reclaims space from all that has been abstract and nonmaterial in modernity. Amid a consumer culture that emphasizes the body as the bearer of cultural symbolism, and therefore an unfinished entity, body art appeals to an enormously broader audience than other sorts of provocation-art.[10] Meanwhile, the academy propagates theories of the embodied subject.[11] Whether this trend constitutes indulgent degeneracy, psychological profundity, or simple nostalgia becomes the crux of aesthetic debate.[12] The critic Robert Hughes once satirized this movement: "You don't like my warm guts? Yeah, you and Jesse Helms, fella!"[13] But as in computing, so in art; some paradigm shift has occurred. As insider-art spokesman Hal Foster declared, "This shift in conception—from reality as an effect of representation to the real as a thing of trauma—may be definitive in contemporary art."[14]

In the study of human consciousness, will the already much-mourned "loss of the subject" be consummated by the blurring of distinctions between the body, technology, and symbolic reasoning? Those are questions for our best philosophers.

Embodied Being

"I refute it thus!"
—Samuel Johnson to Boswell, kicking a large stone in response to their discussion of Bishop Berkeley's proposal of the nonexistence of the material world.[15]

Our inquiry into embodiment must acknowledge some basic philosophical principles. Without delving into this very far right now, note that the mind-body problem has been at the heart of philosophical inquiry for a very long time. For instance, in the Bible, Romans 8:5 says "For they that are after the flesh do mind the things of the flesh; but they that are after the Spirit do mind the things of the Spirit." It is also central to western enlightenment philosophy of the seventeenth and eighteenth centuries. From the discourses on disembodiment that so offended Boswell and Johnson, into twenty-first century theories of cyberspace, a dualism of mind and body has dominated western thought. For some direct evidence of how deeply contemporary thinkers still live with a mind-body split, just watch them act it out by reading the *Wall Street Journal* while they work out on the Stairmaster at the gym.

Descartes had famously asserted that an independent spirit, which inhabited the body, was the impetus behind mental states. The body merely belonged to a world where organic forms operated under the same laws as the rest of nature; that is, mechanistically and without higher goals.[16] This dualism placed disengaged thought ahead of embodied action.

It also led to strong notions of *a priori* space. Preexisting space was the means by which relations among objects could be realized. One consequence of this world view has been a corresponding dualism of culture and nature. In this understanding, a culture imposes symbolically constructed categories on neutral, preexisting nature. An eighteenth-century formal garden expresses this view, for instance. So does a twentieth-century mentalist theory of cognition. An information-processing model of mind assumes a detached subject who is constructing—and then imposing—mental representations. These constructs interpret stimuli from a physical world, but that world is neutral.

Internet users reenact this concept. In believing that they "visit" sites when in fact their browser software downloads packets of data to wherever they are sitting, people suspend disbelief about disembodiment. This aim is credible enough. From ancient Vedic and Platonic philosophies to medieval aesthetics to modernist utopianism, more contemplative souls have always aspired to rise above the mud, disease, and battle that have been humanity's physical lot. People have always quieted themselves, and in some sense left where they are, to contemplate, deliberate, and imagine.[17]

There would be no need to raise all these points if recent brain science had not challenged them. A growing consensus among the biological-naturalist camp of cognitive scientists contends that mental activity is just as much a biological process as, say, digestion. This view has the significant ramification that a great deal of thought is preconscious—and none of it is dematerialized. Mental attributes and constructs are emergent, much as water is wet.[18] Thus the structure of embodiment, itself a product of adaptation to environment, may underlie emergent intent.

In summary of this new understanding, George Lakoff and Mark Johnson boldly declared: "The mind is inherently embodied. Thought is mostly unconscious. Abstract concepts are largely metaphorical. These are the three major findings of cognitive science. More than two millennia of *a priori* philosophical speculation about these aspects of reason are over."[19]

Already well known in interface design circles for their past work in metaphor, Lakoff and Johnson have now presented a more complete theory of cognitive background. Bodies shape conceptual structure; environmental experience grounds metaphor; and a lot more thought is metaphorical than has been assumed previously.[20] Among other results, this leads to an understanding that "The environment is not an 'other' to us."[21] This argument is part of a larger shift that places humanity back within the natural order.

This most recent chapter in the history of embodiment justifies our excursion into pure philosophy before turning to the technological topics at hand.

Mental Models

While acknowledging larger philosophical questions, the discipline of interaction design tends to focus on the mechanisms of perception. For a long time, a cognitive dualism has underlain behavioral approaches to the design of technology. Now some residual connotations of analytical behaviorism must be overturned.

To begin, there exists a claim that only humans have a conception of the world as it is from no particular standpoint.[22] Wittgenstein said that a cat can find its way around the neighborhood—but that it cannot see itself finding its way around the neighborhood. To do the latter would require a reflective "survey perspective" that appears to be distinctly human. For an example of such a perspective, to count the number of windows in your house, you do not have to be in your house. To recognize your house in an aerial photo, you do not need to have seen it from that orientation before.

"A disengaged picture of a persistent world" is the basis of a spatial mental model, which is a principal category in human thought and which remains a fundamental issue in philosophy and cognitive science.[23] Apparently humans assimilate their surroundings by means of mentally constructed representations of spatial relationships. Formerly, researchers held that such environmental schemas are purely mental, but now there is greater recognition of direct engagement and peripheral awareness as complements to deliberative mental models.[24]

In comparison with overt behavior, peripheral awareness tends to be more difficult to study in controlled experiments. Tacit knowledge loses something in the translation to conventional external representations. Understandings based in activity cannot always be articulated without stopping that activity.[25] Where modern researchers confined themselves to behavioralism in the name of scientific certainty, a limited version of environmental psychology emerged.[26] Thus spatial behavior has a well-developed body of scholarly findings, yet our knowledge about shifts in intentional frames of reference is less certain.[27]

For example, many of the most prominent studies of spatial mental mapping have examined the readily documentable process of wayfinding. The Siegel and White studies of 1975 established the distinction of route and survey perspectives, as well as the use of information processing in wayfinding.[28] Much subsequent study has reinforced the view that navigation consists of making decisions at landmarks, even if the resulting "picture" is less of a map than a recombined collage.[29] This, too, predisposes researchers toward the topic of wayfinding, for it turns it into a problem in information processing.

Architects and planners explored cognitive mapping a generation ago. The pioneering work of the urbanist Kevin Lynch is known to many technology designers forty years later.[30] Following Lynch, academic enthusiasm over mental maps of built environments perhaps reached a peak in the early 1970s. Then as psychologists found limits to geometric coherence, and architects found some of their essential understandings unquantifiable, research interests moved on.[31]

In their reductionism, the first generation of findings on environment and behavior have left out two particularly vital concerns. The first of these is *intent*. Intentionality counters behavioralism with a concern for attitudinal or perceptive states that need not result in overt actions, or that at least precede actions. For example, the act of walking down a street may be shaped by what one is looking for, whether one is in a hurry, or whether one feels well dressed.

The second omission is *context*. Contexts do not induce actions so much as shape perceptual selectivity, provide background cues, and enable the application of tacit knowledge. Active embodiment cues what would otherwise be isolated sensory awareness. Intent in context causes cognition to be about something. Here begins an interest in pre-dispositions.

This shift begins with the principles of phenomenology.[32] "Theory of the body is already a theory of perception," wrote the philosopher Maurice Merleau-Ponty. "Our own body is in the world as the heart is in the organism: it keeps the visible spectacle constantly alive, it breathes life into it, and it sustains it inwardly, and with it forms a system."[33] Atop a continually changing substrate of embodied perception, the abstract mental model arises only occasionally, and only when necessary.

> The body is our general medium for having a world. Sometimes it is restricted to the actions necessary for the conservation of life, and accordingly it posits around us a biological world; at other times, elaborating upon these primary actions and moving from their literal to a figurative meaning, it manifests through them a core of new significance: this is true of motor habits such as dancing. Sometimes, finally, the meaning aimed at cannot be achieved by the body's natural means; it must then build itself an instrument, and it projects thereby around itself a cultural world.[34]

Such phenomenology challenges the presumed neutrality of mind-body dualism, on the grounds that the objects have universal essences.[35] The embedded essence affords intuition to subjective intent. In other words, repeated encounters with objects in contexts let us become aware of those objects before any conscious deliberation about them and, furthermore, affects what is likely to rise to consciousness. Dogs are especially inclined to see other dogs, for example. This appears to be a fundamentally structuralist approach; the object's reality becomes understood structurally through the accumulated experience of its many possible instances. It is also constructive; a phenomenon is the moment when the intuition grasps an essence.[36]

Of particular interest to interaction designers today, phenomenology has provided a more practical approach to cognition.[37] Heidegger held that we understand the world in terms of what we can do with what we find of it. Merleau Ponty described how innate structures precede modeling and making. The psychologist J. J. Gibson extended these undertandings to a focus on interaction. In his landmark work, Gibson laid a foundation for understanding human-environment interfaces.[38] His concept of "affordance," now so often abused, interpreted the world as an offering of perceptible structures of possible actions, which are grasped through engaged and not necessarily deliberative action. This is chiefly a claim for direct perception. Here seeing and knowing are one. "Seeing as" combines vision, embodiment, and environment. Haptic orientation shapes this seeing. This continual, preconscious condition underlies, and does not always require development into, discrete mental constructs. This means that learning is a lifelong process that takes place largely in the background.

Embodied Learning

Under larger philosophical questions of intentionality in contexts, the search by interaction designers for a practical means of creating usability tends toward issues in learning.

Contextual learning begins at embodiment, remains largely personal, and is lifelong. A newborn infant may not even know he or she has a body—only needs—and therefore may not be able to distinguish between self and environment.[39] From this limited and very egocentric frame of reference grows an increasingly articulated understanding of an outside world. Because contexts are learned through actions and events, much of this understanding is based on memories of interactions: object permanance, landmarks, proportional configurations, spatial categories, procedural contexts, swapped frames of reference, geometric measures, building elements, generative typologies, systemic behaviors, formal elegance, regional characteristics, ecological sustainability.

Understanding proceeds with a constant cycle of construct readjustment.[40] Environments that subtly challenge our constructs provide more satisfaction than those in which everything conforms to expectations. In learning one does not simply form a picture of a static world, but instead actively shapes that world according to emergent understandings. It is important to note that embodied learning occurs at several levels, ranging from the preconscious engagement of affordances, to the personal construction of mental models, to the cultural mediation of spatial literacy (figure 2.1).[41] Etiquette is an example of the latter. For instance, a new arrival on a university campus goes through a period of willingness to enter any building socially, and then a period of settling into a routine of places, before developing a subtler use of the campus.

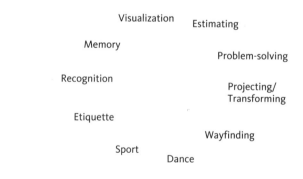

2.1 Aspects of spatial ability

Throughout these levels of learning, engaged interaction is at least as important as detached perception. The possibilities of the world correspond to the capacities of the body. Those capacities may be innate or acquired, direct or mediated, universal or constructed by a particular culture.

This points toward the roots of interactivity. Contexts are full of props and cues, which serve as learning resources and memory devices for evolving patterns of usage. Many such cues serve as constraints; context rules some things out so that others may receive closer attention. Those perceived resources are appropriated toward an active intent.[42] This grasp is engaged but not necessarily reflective. It is as much a product of the abilities and intents of the subject as of the properties of the object. This is one reason why the use of tools transforms the perception of environment. This actively engaged way of learning about the world challenges the assumption that technology is a purely symbolic literacy, independent of ground.

Spatial Literacy

Studying interactivity reveals the cultural aspects of embodiment as well. More of us note how cultures and their symbolic systems mediate learning processes. In this regard, it becomes possible to speak of spatial literacy.

For example, one learns to read a city without the aid of books and maps, and to do so partly on the basis of experience with culturally similar cities; some individuals and some cultures develop more ability than others at this.

As with most cultural differences, spatial dispositions show up in language. For example, while an Englishman might live "in" a street because a street was once a public living room, an American lives "on" a street because it is merely an address, and to judge by current naming practices, would prefer to live on a road, a lane, a court—anything but a street.

Language itself plays an important role in spatial literacy. Language abounds with bodily metaphors that recall the experience of environment. To return to the case of the corner office, note that the word *corporate* derives from *corpus*—the body. Within business banter, consider the prevalence of an upward schema, which is the direction the body grows. Thus, for example, in a corporate setting you might hold high ambitions, dress up, rise to the occasion, stand up to authority, or do some heavy lifting.

Spatial language builds from words to metaphors, narratives, and even world views. Mythological narratives color local landscapes.[43] Allegory often employs spatial progress (e.g., a pilgrimage to the inner circle). Epics construct detailed and coherent worlds. Thus a culture uses spatial configuration as a memory device. For example, the New England village persists as a representation of a particular spirit. Its water mill, town common, meeting house, and hub of roads in from neighboring villages provide a cultural memory of life before steam plants and railroads. Its building forms also provide civic legibility— you can tell where to go to find public life.[44]

Spatial literacy should not be confused with literal signage declaring space. They are quite opposite. Whereas an outsider who lacks local spatial literacy needs the latter for guidance, a literate denizen reads a space from its events and its symbols, like animal scat on the trail, and does not enjoy being told where to turn, what exactly occurs in each place along the road, or that a brand-name experience will protect him from unwelcome surprises.

Social Configuration

Social territories involve a literacy also. The cultural geography of belonging and identification depends on learned spatial cues that are not necessarily hard-built in an explicit form. Moreover, as Foucault insisted, the existing outward forms afford readings of social relations that their owners would sometimes prefer to remain tacit. And this literacy is bodily; power and discipline become readable in their conventions for organizing bodies.[45]

Set up a group of animals in a fenced-in area and soon individuals will have staked out their territory and their pecking orders, all of which will be clearly expressed in a settlement pattern. Patterns of spatial usage tell us as much about a species as the anatomy of its individuals.[46]

Humans are no exception in this regard. Put a group of people in a room, and they will quickly organize themselves. Consider the importance of social distance, presentation of self, and territoriality. On a larger scale, note collective memory and the anthropomorphizing image of the city.

Interpersonal distance is the great mediator of social standing. On almost any scale, the inflection of interpersonal distance provides a tacit set of social cues. This is important to natural interactions; the tacit geography of these social relations constructs place.

Social distance thus establishes categories of experience, from the intimate to the collegial to the public.[47] Framing the interplay of embodied behaviors remains the most important function of environment. Building instrumentalizes and civilizes social distance. Architecture consists of built social relations. Its behavioral framing establishes who may see whom and under what protocols. Systems of social distance become more elaborate in wealthier and more sedentary cultures, that is, those with an investment in fixed places.

Body image reinforces these systems with distinct codes of behavior and dress. These etiquettes do not stifle social expression so much as specialize it. They do not fix distance rigidly so much as establish the variables for fair play of the game.

In a favorite example among environmental psychologists, a sense of crowdedness depends on what people are doing.[48] Concertgoers have a different notion of personal space than ballroom dancers. Labor-intensive rice farmers pack together more comfortably than capital-intensive, industrialized wheat growers. Subsistence hunters may feel crowded in the wilderness when food gets scarce. In the city, variations on crowding make life enjoyable. Nightclub mosh pit enthusiasts go for the contact that they lack while sitting at computers all day.

Note that habitual embodiment in a persistent environment of quality lets adults play too—at daily games of social standing. As an example of sophisticated play, consider the case of the evening *paseo*, the traditional Spanish custom of strolling back and forth along the town square. This arrangement gives each citizen a chance to present himself or herself to the community. Compared with the *paseo's* standards of social skill, many modern Americans do not walk very well.

Besides these finer exercises of social distance, territoriality establishes elaborate patterns of enclosure and access. Crude territorial marking underlies elaborate form, and built space is as much a display of ownership as a framework for social conduct.[49] Great value is conferred on the site of dwelling. By establishing a center outside the body, to house, rest, and reflect the body, the form of dwelling reflects the condition of embodiment more directly than just about any other social construction. Dwelling is grounding, in the oft-cited Heideggerian sense.[50] To dwell is an intentional state, and a historical one, in which one identifies with a place.[51]

Note also that just as enclosure indicates ownership and regulation, so open space equates with freedom. Open space is room to move and grow. Space that is open, yet owned (with enclosures that are intangible or removed from sight) is the best of both worlds. In America, especially, such space expresses prestige.

Cultural Disposition

Entire cultures dwell, and they build stories and literacies around that fact. As evidenced by cultural differences in land use, for example, cultural bias develops at the level of the built environment. In traditional societies, tales beginning from the center present the universe as an orderly and harmonious system, which settlement patterns attempt to reenact. Perhaps the first stories described the best routes to hunting grounds. Then came metaphor: to grow *up*, to form an *out*look, or to dig *in*, was to remember space. Indeed most narrative imagery and allegory was somehow grounded in common spatial experience.[52] From this chthonic basis, each culture could built its own orientation. Landscape features acquired personalities, geographical excursions reenacted histories, social correspondences were applied to the cardinal directions.

The anthropologist Mircea Eliade once described such spatial attributions according to the principle of homology.[53] In his description, homology is a formal similarity between a sacred condition and a profane reality. It gives meaning to the human condition by repeating the structure of spiritual belief systems in the configuration of the physical world. It grounds. Such spatial relations deepen a sense of connectedness, orientation, and duty to the land. Some contemporary cultures perpetuate this quality more than others. For example, the Balinese maintain a mythology built from the fact that on their island a striking number of rivers flow in parallel from volcanoes in the north to the sea in the south. In the resulting moral geography, north/uphill/sacred is where one can pass from one cosmic condition to another, and south/downstream/profane is to be avoided.[54] Hence the latter area has been more readily conceded to westerners, who enjoy the surf.

Quite often a people forgets that it has a particular environmental orientation, however. A culture may not always acknowledge that even the most mundane environmental configurations are far from inevitable: choices have been made. Recollecting this is what makes travel so interesting. More than beaches, what attracts western tourists to Bali is how everyday space manifests divinity. But the use of everyday space is also full of choices. A tourist seeing some other culture using, say, fences so graciously to facilitate close living may suddenly realize the arbitrariness of that peculiar American preference for the open, nominally democratizing, expanse of lawn.

As each culture develops its own environmental ordering as a foil to the world's indifference, settlement patterns not only reflect but then also shape beliefs (figure 2.2). As cultures become identified with their

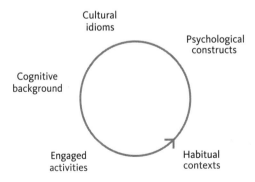

2.2 A general notion of construct adjustment in habitual contexts

peculiar spatial customs, landscape tends to serve as the best framework for narrative memory. Thus Cicero could write: "*Quacunque enim ingredimur, in aliqua historia uestigium ponomius;*[55] "For walk where we will, we tread upon some story."

Deskilling

Unfortunately these patterns can grow too rigid. When particular arrangements have proven convenient, or have been socially or cosmologically conditioned, or have too often been validated in individual experience, they become less flexible. This inflexibility reinforces cognitive preferences. Those events that reinforce the schema appeal more than those that challenge it. Preference becomes predisposition; constructs become imposed on environments, and challenges to them get ignored, at both the personal and the social level.

Thus the cycle of embodied environmental literacy can turn downward. Technological convenience allows many helpful new constructs to form, but it also allows events that would normally serve environmental learning to dwindle. Quite often such troubles are blamed on cultural considerations such as economic models, but personal considerations such as body image also contribute. In this view, if only more people could make the connection between bodily schemas, domestic patterns, city form, and regional identity, the world would be in better shape.[56]

Technology design too seldom taps latent predispositions (skills we already have) and too often requires arbitrary instruction (still more skills we must learn). Technology has often extended life experience beyond the scope of bodily schema. It has shifted organization from space to time. As societal forces become higher dimensional and less directly visible, three-dimensional spaces of experience seldom remain coherent. Much about modern life frustrates our body image imagination.[57] All this produces distress.

Even the built environment discourages the full exercise of embodiment. Writing in the 1970s, long before the age of virtual reality, the architects Kent Bloomer and Charles Moore cautioned against deskilling in a culture based on visual novelty: "One of the most hazardous consequences of suppressing bodily experiences and themes in adult life may be a diminished ability to remember who and what we are..."[58] According to the ecologist Wendell Berry: "We have given up the understanding—dropped it out of our language and so out of our thought—that we and our country create one another"[59]

On a subdivided farm in central Massachusetts, a developer puts up Yankee saltbox houses. This traditional two-story building type is named for the way its roof comes down to the first story on one side to fend off the winter wind. But the developer sites these saltboxes arbitrarily, and many of the roofs end up facing the sun, not the wind. "Man dwells badly," wrote Le Corbusier, "and that is the deep and dear reason for the upheavals of our time."[60]

The Case for Ground

Although some larger critique of settlement patterns and the spatial deskilling they reflect seems inevitable, our present concern is much simpler. If spatial deskilling has emerged as a major problem of our time, then our technological constructs must be adjusted to confront this. The appropriate technology will be that which taps into and uses embodied predispositions. Amid present movements toward pervasive computing and situated interaction design, we need to base a theory of context on these many principles of embodied environmental perception.

Environmental predispositions exist. Abundant evidence for this condition can be found in a comparison of settlement patterns, in the nature of recreations, and in the scholarly study of language and thought. The relevance of these patterns and practices is demonstrated by one of the fundamental tenets of scholarly inquiry, namely that independent social productions reveal common underlying structures. It is also shown by the breadth of mainstream media focus currently given to problems in this area, particularly with respect to the socially limiting configuration of built and electronic environments.

The main objection to this argument is that any sense of place is highly personal and very difficult to measure. So too are hope, faith, and happiness, of course. Design, technology, and academic inquiry cannot afford to continue to ignore human emotional and intentional states (rather than merely human behavior) simply for the sake of certainty. Compared with some rather more difficult social conditions, attention to embodiment provides a fairly straightforward opportunity to develop the expression and valuation of properties that for too long have been dismissed as unmeasurable.

Another major objection concerns that fact that environmental sensibility cannot always be advantageous. We cannot always stop and smell the roses. Overattention to the periphery may distract from urgent decisions to be made in the foreground. When crossing a street, you do not have time to study the surrounding scenery; you must get to the other side.

In the end, people prefer to operate on a full spectrum of focus, from deliberation to contextual association to the unconscious application of cognitive background. Qualifying the value of environmental knowledge according to this spectrum is not so much an objection as a way toward the better design and practice of appropriate technology. Embodiment is a property of interactions; latent embodied abilities exist; and good interactive technology lets us exercise these abilities.

3   Habitual Contexts

Not all is flux. Much as a river needs banks unless it is to spread aimlessly like a swamp, the flow of information needs meaningful contexts. Even in an age in which distance has been annihilated, location still matters.

The built environment organizes flows of people, resources, and ideas. Social infrastructure has long involved architecture, but has also more recently included network computing. The latter tends to augment rather than replace the former; architecture has acquired a digital layer. As with past layers of technology, such as electrification, mechanical equipment, and transportation, so now digital technologies extend architecture's reach. In doing so they take advantage of architecture's duration. The older and more persistent the grounding structure, the more likely that it has shaped environmental predispositions. In contrast to more ephemeral electronic works that compete for the momentary attention of casual viewers, built environments act as enduring background, and their design is directed inward toward their regular inhabitants.[1]

There, in our most habitual contexts, embodiment provides a continuing basis for human-centered design. For much as the body imposes a schema on space, architecture imposes a schema on the body.[2] The proportions, image, and embellishments of the body are reflected in the proportions, image, and embellishments of buildings. Similarly, cities reflect the form of their buildings, cultural landscapes reflect the structure of their cities and towns, and mythologies orient all of these in the world. Although the sciences have extended this scale of artifice farther into the immense and the microscopic, the orders of magnitude nearest to human dimensions still affect everyday experience most directly.

The disciplines of architecture and interaction design both address how contexts shape actions. Architecture frames intentions. Interactivity, at its very roots, connects those mental states to available opportunities for participation. These processes are ambient. Their benefits are to be found in the quiet periphery, and not in the seductive objects of attention. Why this is so was put well by one of architects' favorite thinkers, Walter Benjamin, who reminded us that "architecture is experienced habitually, in a state of distraction."[3]

In turning from embodiment in person to embodiment in the built world, it will help to define some terms. To begin, let "setting" describe objective, a priori, space. "Context" is not the setting itself, but the *engagement* with it, as well as the bias that setting gives to the interactions that occur within it. "Environment" is the sum of all present contexts. According to the cognitive principles laid out thus far, environment is not an other, or an empty container, but a perception of persistent possibilities for action.

"Space," like embodiment, has occupied philosophers from the ancients to the latest wave of cyberpunks.[4] Because it allows motion, space has been intrinsic to modernity. Space is a means, and not a mere setting, at least according to the philosophical traditions charted by Kant. It is the form of external experience as distinguished from the things encountered within that experience.

"To speak of 'producing space' sounds bizarre," wrote the critical theorist Henri Lefebvre in 1974, "so great is the sway still held by the idea that empty space is prior to whatever ends up filling it."[5] Notions of preexisting space now give way to emergent phenomena. Wherever goods, people, or electronic communications flow, spaces form around them. This emergence has been particularly evident in the case of disembodied electronic channels. In what the sociologist Manuel Castells named the "space of flows," global capital has apparently invented a new kind of space for itself—one whose spatiality emerges from, rather than preceding or containing temporal activities. But as Castells explained, this net changes relations between physical places more than it does away with them. "The space of organizations in the informational economy is increasingly a space of flows.... However, this does not imply that organizations are placeless. On the contrary, we have seen that decision-making continues to be dependent upon the milieu on which metropolitan dominance is based; that service delivery must follow dispersed, segmented, segregated markets.... Thus each component of the information-processing structure is place-oriented."[6]

Places emerge at crossovers between infrastructures. Where one flow prompts, regulates, or feeds another, development occurs. Where the boats met the trains, great cities grew. Increasingly, such connections occur between digital and physical infrastructures. Electronic communication has intensified, not undermined, the hubs of activity in the world's entrepots. This intensification is reflected in the current practices of urban design. As cities everywhere move to correct the separation of use wrought by the industrial age, we have rediscovered how the flows of people, goods, and information are most valuable wherever they are most closely intermingled.

In movements we have seen described as "after cyberspace," information technology contexts are no longer valued for immersiveness so much as for "periphery." Information technology design has occupied itself with tools for deliberative reasoning—a process that occurs in the foreground of human attention. In a recent standard text on interface design, Apple Macintosh project creator Jef Raskin emphasized the term *locus* of attention. "We cannot completely control what our locus of attention will be.... For our purposes, the essential fact about your locus of attention is that there is but one of them. This underlies the solution of numerous interface problems."[7] Unfortunately this attention remains finite while the number and complexity of tools continues to increase. In what has become a problem for almost all design disciplines, the foreground is full.[8]

In response, most agendas of physical computing share a belief in "periphery." As defined by John Seely Brown, the former director of the open research center Xerox PARC, "periphery is background that is outside focal attention but which can quickly be given that attention when necessary."[9] This is one way to deal with information overload. "Periphery is informing without overburdening."[10] Trying to keep too much in the locus of attention tends to be stressful. We find it more natural to use our considerable powers of sensing the surroundings, and then to experience more capacity and resolution where our attention is focused. Thus, as Brown observed, bringing something back from the periphery to the center of attention is a fundamentally engaging and calming process.

Pervasive computing takes this approach beyond the information context to include physical architecture. Graphical user interfaces have long been built on principles of shifting focus—picking up a tool, opening and closing a window, etc.—but they still leave us staring at a cluttered screen. Portable and embedded systems take the information processing out into the physical realm, where the capacity for periphery is deeper and the act of bringing things to the center is more intuitive. For example, tagging systems can mark parts inventories for direct use by hand-held devices without recourse to a desktop database. Principles of periphery can help reduce ·ontention on a screen, of course, but they also suggest a larger shift in our goals for natural interactions.

This is mainly a matter of embodiment in context. Our embodied predispositions have been underfed while our foreground deliberative attention has been oversaturated. To change that balance, we need to change what we expect of interactive technology, and where we expect to find it.

Context and the Roots of Interactivity

As reflected by so much recent emphasis on embodiment, contextual factors matter more than early researchers in interactivity anticipated. If more recent study finds the phenomenology of engagement at the roots of interactivity, it is because these designers build technologies around everyday life. This shifts design values from objects to experiences, from performance to appropriateness, from procedure to situation, and from behavior to intent.

With its new emphasis on intentions in activity, contextual design departs from an earlier generation of inquiry into environment and behavior. Whereas that work aimed to reduce design to a linear, predictable process, based on measurable models of conditioned response, the current work recognizes the importance of expectations.

"When we speak of 'direct manipulation,' 'intelligent agents,' 'expert behavior,' and 'novice behavior,' we are really positing concepts in which consciousness is central," the anthropologist Bonnie Nardi has explained.[11] Intent makes people different from machines in any flow, and it gives an asymmetrical cast to the relation between people and things (figure 3.1). Cognitive science has emphasized mental representations at the expense of context. "Thus we have produced reams of studies on mentalistic phenomena such as 'plans' and 'mental models' and 'cognitive maps,' with insufficient attention to the world of physical artifacts."[12] Designers more interested in rich description than in predictive models tend to welcome such emphasis on artifacts. As a way of describing the intrinsic unity of context, activity, and intentionality, "activity theory" has become a useful expression.

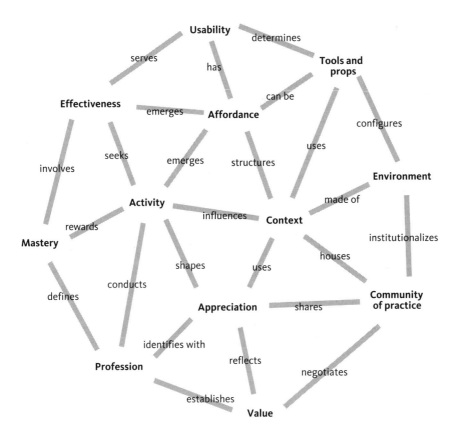

3.1   A concept map for embodied activity in context

The word *situation* keeps us mindful of the ethnographic perspective. Ethnographers remind us that actors play their settings. An improvisatory action grows out of the immediacy of a context. Situated action theory explains how experts engage contexts. As voiced by the work practices ethnographer Lucy Suchman, who introduced the theory into a still very mechanistic field in the late 1980s, "The organization of the situated action is an emergent property of moment-by-moment interactions between actors, and between actors and the environments of their action."[13] Within the situated action model of work, actors operate within a stable institutional framework, or "arena," to create personally ordered versions of the environment matched to their respective habits and goals.[14] Habitual contexts support courses of action in which effectiveness has been internalized enough that it need not rise to the level of a conscious mental model. For example, a competent intern makes hospital rounds according to well-established procedures, but an expert doctor makes his or her rounds according to more tacit and personalized criteria.

Many processes of everyday life involve such sensibilities. For example, a resident who walks through his or her neighborhood on the way home from work casually notices incidental changes to objects and surroundings, and these may prompt improvisatory shifts of intent about what to do that evening.

Persistent Structures

The more enduring the environment, the more it shapes our expectations without saturating our attention. The phenomenology we have examined suggests the need for more design emphasis on lasting backgrounds.

Designers do seem to understand the importance of contextual perceptions. This has been demonstrated by the overexposure of Gibson's word *affordance*. A coupling of perceived resources to active intent creates a context for that action.[15] The sum of all such contexts present is the environment. One's active state heightens this impression. Thus affordances are inherent properties of environments. When affordances are perceived similarly by different people, the identity of the environment is reinforced. Sensibilities shared longer or more frequently become the basis of place identity.

As people learn from their settings, they come to associate them with particular states of intent. Intent is important because it causes engagement of context to be, as Nardi put it, "about" something.[16] This is why meditation teachers insist that a particular spot in the house be set aside for no other purpose.

Habits matter. Accumulated experience of intent and action allows more abstract mental models to develop. It is especially in habitual context that perceptions of affordances, persistence, and situation emerge.

These ideas about activity in context have become fundamental to the field of interaction design (figure 3.1). Although they are vexing to some researchers because they are difficult to measure, they are encouraging to architects, or to those who understand the design technology as architecture, because they make it easier to raise questions of cultural identity.

Developments in pervasive computing amplify the importance of these arguments. "At the heart of tangible computing is the relationship between activities and the space in which they are carried out," interactivity scholar Paul Dourish has observed. "Tangible computing expands this in three ways: through the configuration of space, through the relationship of body to task, and through physical constraints."[17] Embodiment is not just a state of physicality, but a situation in real time and place.[18]

Henceforth the design of information technology expands its subject from artifacts to their contexts. Through context, designed objects become expressions of identity, signs of differentiation, tokens of communication, and a natural support for relationships.[19] The production of environments, organizations, services, and communication among all these becomes the domain of design.

Scale and Pattern

The habitual experience of an activity employs an ever-increasing flow of symbols and inevitably occurs in some setting. Fixed configurations of spaces, props and artifacts thus support activities in ways that go beyond housing them.

One important perception of how well such support fits an activity is scale. To continue an earlier example, consider the perception of crowding. As should be of interest to interaction designers, a feeling of crowdedness depends entirely on what people are doing. To a writer, just two people can be a crowd, but at a nightclub, the more people packed in, the better. Differing numbers of people in a fixed space change the scale of an event. Changing the space that a fixed number of people use to do something also alters the reality. Bigger is not necessarily better in this regard. Ten people can meet more comfortably in 400 square feet than in 4000. Indeed, quite a sensitivity exists in this regard. A difference of 100 square feet could have an influence on the tone of a ten-person meeting. For more sedentary activities, matters of scale can be more fine. A difference of an inch in the space between rows of airplane seats can change the experience of flying.

As a consequence, built environmental dimensions are not arbitrarily scalable. Consider how, in a family kitchen, the scale of the countertops is related to the scale of the room. A small room might have only a straight counter along one side. If it is slightly wider, there is space to turn a corner with the counter. If it is wider still, there is room for an island. Even if one were able to transform this width dimension continuously, discrete breaks from one arrangement to another would occur. This simple example demonstrates the importance of types.

Living systems tend to maintain rich interrelationships of scale. Popular notions of ecological design emphasize this. Similarly, the appeal that draws tourists to traditional landscapes and towns has much of its basis in a complex interweaving of scales. In contrast to the monoculture of suburbs designed for the automobile, the center of a European city offers narrow streets and lofty monuments, grand boulevards and intimate courtyards. Traditional areas of the island of Java, one of the most densely populated places in the world, retain more appeal than much less densely settled places in the automobile-centered world, simply because the settlement patterns are intensive rather than extensive. That is, they make much more sensible use of intimate scale.[20]

Repeating relationships embody workable conventions. These are not rigid rules, but transformable configurations. This phenomenon was documented in the perennially debated work of Christopher Alexander, *A Pattern Language*.[21] This language of some 250 architectural elements was mainly addressed to the affordances of built space for living. For example, walking under an arcade around a courtyard helps an otherwise too-sedentary person think something through. The space is appealing because it provides a seasonal and psychological transition between indoors and outdoors. A courtyard that opens out to a view in one direction contrasts its closure with that openness, and releases the space from any possibility of claustrophobia. Alexander emphasized situations too. "The life of a house, or of a town, is not given to it, directly, by the shape of its buildings, or by the ornament and plan—it is given to them by the quality of the events and situations we encounter there. Always it is our situations which allow us to be what we are."[22] Livable contexts do not occur in endless free form so much as they establish a persistent and smallish set of types. "Nothing of any importance happens in a building or a town except what is defined in terms of patterns which repeat themselves."[23] "And what is most remarkable of all, the number of the patterns out of which a building or a town is made is rather small."[24]

Typological Abstraction

A theory of place for interaction design incorporates embodied cognition into a workable design philosophy through types (figure 3.2). In a single design notion, type unites periphery, passivity, phenomenology, adaptability, affordance, facility, appropriateness, and scale. Thus it is a difficult term, which means different things to different disciplines. For present purposes, consider type not as a mere functional classification, but as a generative design abstraction. This is a central idea for more context-based pervasive computing, and it should help bring interaction design into closer relationship with architecture.

A type may be as much about form as function. For example, the town square is a distinct urban type more on the basis of its intrinsic

3.2 A concept map for typological design

form than on its uses, which may vary from week to week, and century to century. Indeed the square is memorable for how its configuration affords so many uses, the accumulation of which increase its resonance as a type.

The psychological importance of type is not only a consequence of activity in context, but also of cultural identification. This is a foundational notion because enduring structures become repositories of human, cultural, and organizational capital. The patterns of activities first become means toward those activities, then come to suggest and represent them. Stability, duration, fixity, and repetition are all qualities of experience by which human contexts acquire value. Arrangements for the fixity of cultural proceedings are generally known as institutions.[25]

As the anthropologist Mary Douglas succinctly put it, "institutions confer identities." Physical sites embody expectations. "The whole approach to individual cognition can only benefit from recognizing the individual person's involvement with institution-building from the very start of the cognitive enterprise."[26]

The casting of daily life into particular scales of form reflects lasting social agreements about categories and values. Here is the inevitable "social construction." This is how an institution can be said to have scale. Through not only its endurance, but also the way it manifests convention, built form provides identity for both an association and its constituents.[27]

Such value is especially evident in cities. A livable city is made up of types. Some of these, such as the sidewalk café, become valued for all the experiences that have accumulated there. Any institutionalization is purely unofficial. Other sites declare values and expectations more deliberately; a public library does this well, for example. Cultural distinctions in handling these typological elements become sources of exchange and identity in themselves. Particular places are known for their types. New Orleans has its patios, and Brooklyn has its brownstones.

If types are a way of representing livable arrangements, it is because they help make the link from body to building to city to landscape to universe. From this basis, Douglas was able to argue that types are "foundational analogues," and moreover that these correspond to archetypes of living systems. "There needs to be an analogy by which the formal structure of a crucial set of social relations is found in the physical world, or in eternity, or anywhere, so long as it is not seen as a socially contrived arrangement. When the analogy is applied back and forth from one set of social relations to another, and from these back to nature, its recurring formal structure becomes easily recognized and endowed with self-validating truth."[28]

Architectural Type

Architecture demonstrates these principles particularly well. As an enduring framework that is used habitually, architecture provides an obvious basis for a more context-based approach to interaction design. Architecture surpasses most other technological productions at institutionalizing spatial arrangements to the extent that they shape cognition. A culture's perennial spatial forms perpetuate a particular cognitive background. This is why one of the best criteria for appreciating architecture is whether it is memorable.

Architecture serves the body and not just the gaze. It is not just perceived, but inhabited. At a mundane level this means that architecture includes such things as stairs, toilets, and heating systems; but at a more imaginative level, this embodiment gives architecture the power to order the external world according to our inner topographies.[29] When this anthropomorphizing function is working, architecture helps us move from personal to societal scale.

Like the human body itself, an architectural type supports endless variations whose appeal is in their subtlety, not their shock value. As in people's faces, we appreciate variations on architectural themes mainly in terms of the relationships of familiar details.[30] These details remain in memory better than radical aberrations. The brownstones of Brooklyn are distinct from the marble-stooped row houses of Baltimore, for example, and they are also subtly varied among themselves.

As a generative abstraction, architectural type is essentially morphological. In the case of dwelling, for example, important types range from palazzo to terrace to tent.[31] Particular forms and their variations become distinct to particular cultures and locales. The pleasure of these arrangements is built in part on the richness of their architectural components. Elements such as portals, pathways, canopies, and courtyards represent archetypes too fundamental to be exhausted or ignored.[32] Especially when aggregated into neighborhoods, design that inflects these achieves more humane detail than design that invents alternatives.

Typological design seeks a balance of convention and invention. Too much convention becomes stultifying; too much invention becomes inane. In the review of embodiment and context, we have seen how skills and knowledge build from accumulated knowledge of persistent structures. From this it follows that in the language of design, types enable at least as much as they restrict. Typological design is not a rigid set of rules, but instead a body of phenomenal essences which play themselves out differently in each instance. This inflection provides a richer basis for building workable arrangements than does radical atypological invention.

More so than the purveyors of pervasive computing, architects have learned the costs of world making. Alas, too much of today's built world is the environmental crisis that we can touch. To many of us this crisis has largely been a lapse of typology. Picture a split-level four-plex apartment building squeezed onto one narrow lot on a street full of Victorian workers' cottages. Whether or not the newer building performs better (mechanically, electrically, thermally, etc.) it just isn't appropriate, and so it has a negative impact on the street.

According to many of recent architecture's critics, the most insightful of whom has perhaps been Stewart Brand, the quest for invention everywhere has undermined the value of the built environment. To Brand, "Aspiring to art means aspiring to a building that almost certainly cannot work, because the good old solutions are thrown away."[33] Part of the problem is that architects are taught to seek the anti-typological. Emblematically, "The roof has a dramatic new look, and it leaks dramatically."[34]

Wealthy patrons of signature architects can pay to make new forms work; everyday design failures occur mostly in everyday building. This is especially true at the scale of the neighborhoods. As critics of sprawl and advocates of sustainability explain, monocultural neighborhoods built from single-function building types do not function especially well as larger systems. The relevance of this critique to the relationship between architecture and interaction design is basically a matter of cognitive background.

As noted in our review of embodiment, some fundamental skills and orientations are being lost. Here with respect to habitual contexts, consider one such orientation: cultural memory. The city is a repository; its landmarks, types, and forms are the access structure. Its civic space—so dear to architects—is a design for interaction. That aspect makes it worth recalling Aristotle's assertion that man is an inherently political being, and the city is the best arrangement for realizing that aspect of human nature.

In addition to questioning the advantages of anytime-anyplace technological freedom, interaction designers should question pure functionalism. As architects know, places aren't just locations with assigned uses or trademark formulas. Cities have histories, in which they have been appropriated for this and that. Accumulated experience of appropriations makes people come to identify with places. Built contexts are collective memory devices, and manifestations of collective cognitive background. As long ago as the 1960s, urbanist dissidents such as Jane Jacobs and Aldo Rossi anticipated today's shift toward these ideas. Rossi's book *The Architecture of the City* remains the seminal declaration of typology. It is of interest to the question of interaction design for its critique of what Rossi called "naive functionalism." To begin this argument he drew particular attention to the monumental nature of urbanistic chunks. "In almost all European cities there are large palaces, building complexes, or agglomerations that constitute whole pieces of the city and whose function now is no longer the original one. When one visits a monument of this type, for example the Palazzo della Ragione in Padua, ...one is struck by the multiplicity of functions that a building of this type can contain over time and how these functions are entirely independent of the form."[35]

"Ultimately we can say that type is the very idea of architecture, that which is closest to its essence," Rossi declared. "Naive functionalism ends up contradicting its own initial hypothesis. If urban artifacts were constantly able to reform and renew themselves simply by establishing new functions, the values of the urban structure, as revealed through its architecture, would be continuous and easily available. The permanence of buildings and forms would have no significance, and the very idea of the transmission of a culture, of which the city is an element, would be questionable."[36]

Technological Change

Technological change has reconfigured buildings and cities in the past, but it has seldom done away with them. Because types adapt in response to conditions and needs, technology can contribute to the resiliency of types.

This is familiar enough in architecture. For example, the high-rise office building was made possible by elevators, and less obviously, the telephone. (The latter allowed offices to be separated from factories and docks, and it allowed a large organization to occupy several smaller floors more conveniently.) The elevator transformed the apartment block, as well, both by allowing it to be built higher and by taking over the space traditionally left for a central courtyard, thereby turning the plan form inside-out, from an O to an X. Plumbing also transformed the apartment block. In the ancient Roman form of the type, the *insula*, only the ground floor had running water, and so this is where the wealthy owner lived.

Electrification, in particular, reshaped domestic life, health care, manufacturing, and transportation.[37] The proponents of pervasive computing often draw analogies with this phase of technological history. The usual explanation holds that embedding sensors and microchips in appliances has similarities to embedding electric motors in appliances. Embedded systems let work be carried out where it is most convenient, and not necessarily at some centralized site. In the factory, portable electric motors transformed manufacturing, and made the assembly line possible, which opened up mass production, which created standard building components—and of course automobiles, air conditioners, and so on. Soon embedded systems were taken for granted.

Two aspects of this explanation are mentioned less often. First, electrification produced uncanny effects. Lighting, for instance, reverses the privacy relation between indoors and outdoors at night by making the former visible to the latter. It also reverses the architectural modulation of classical facades by lighting them from below, rather than from above like the sunlight for which they were designed. These are examples of ways in which new technology makes the familiar seem strange.

Second, new technology produces design opportunities mainly in relation to existing technologies. For example, electrification transformed passenger rail traffic by making it possible to light (and thus build) subways and subterranean train stations. Grand Central Station as we know it was a consequence of smoke-free locomotives and electrical lighting.[38] It also changed timekeeping. Daily lives became arranged around train schedules. Tightly timed commutes were intrinsic to the machine age. They were not just a product of train timetables but also a reflection of a will toward synchronization and metering so powerful that Lewis Mumford declared that the most essential machine of industrialism was the clock.[39] Railroads created a remarkable interplay of space with time. In what has become a favorite story among network technologists today, trains with long enough range and high enough speed eventually forced a remake of timekeeping itself, that is, time zones. We now take those for granted, too, but at the moment of their adoption in 1884 the newspapers lamented how "God's time" had been abandoned for (New York Central owner) "Vanderbilt's time."

Schools, stores, libraries, theaters, banks, trading floors, workplaces, and homes endure, but in often radical new combinations, as a result of technological change. "Recombinant architecture," as William Mitchell called the impact of software on building typology, takes apart many of the spatial linkages we have come to expect, and reassembles them into new forms.[40] This is not the first such period of change: industrialism, electrification, and modernity remade classical types according to radical new programs of functionality and resource use. In the classical city, form announced civic aspirations; in the modern city, form followed mechanized function. In the digital city, form must provide the periphery and the ground for our swapping of bits, our multiplexing of activities, and our continued need for an enduring environment.

Building Backgrounds

By acknowledging its foundations in embodied activity in habitual contexts, interaction design becomes a defense of architecture. In contrast to earlier stages of interface design aimed at building attention-saturating virtual worlds, this new paradigm in information technology turns to building physical backgrounds. The more that principles of locality, embodiment, and environmental perception underlie pervasive computing, the more it all seems like architecture.

The way in which a built environment shapes an organization's activity and represents its aspirations now interests information technologists. This is because information technology, like architecture before it, has become social infrastructure. The word *architecture*, which more literally means "master builder," has been appropriated to describe all manner of technological designs that are infrastructural and irreversible, and that cast everyday activity in a particular way. When the configuration of action is not conveniently alterable, where it has arisen from some particular intent, and where it embodies some clear model of understanding, operating, or cultural aspiration, then it is fairly called architecture.

This has already been beneficial to architecture. Interactivity becomes a remedy for architecture, which as a discipline has ignored usability, performance, and inhabitation in its quest for attention-seeking novelties in form. Architecture needs to rejuvenate itself with interaction design.

The newer field of interaction design benefits as well, for example by becoming more sophisticated about environmental perception. It extends, and does not abandon, previous works of place making. It takes advantage of physical contexts as frames and cues for its social functions. It begins to reflect scale and type in its pursuit of site-specific technology, context-aware systems, and location-based services. It shifts focus from technological novelty to more enduring cultural frameworks.

Together, these shifts suggest more emphasis on quiet architecture. As digital technologies surpass their predecessors at expressing the culture of the moment, particularly in its visual aspects, physical architecture is relieved from its struggle to be at the fashionable center of attention, and returns to what it does better in any case, namely the enduring formation of periphery. Would-be designers of pervasive computing environments should consider an architecture of periphery. Architects of built periphery should emphasize the affordances of everyday life rather than fashionable statements in form. Buildings should be valued for their duration. In contrast to information technology's rapid churn of data, devices, and techniques, quiet material permanence seems like a welcome source of calm. Digital systems are then applied to buildings to give them adaptability, and hence more duration.

In this respect, physical building absorbs pervasive computing, like so many other technological layers before it. Since it is transformed by successive layers of technological development, the fixity of the built environment is not absolute. Since it accumulates, it does have fixity relative to any potentially new layer of systems. Some of that fixity is obsolete legacy, and some is useful armature for extensibility. The endurance of these layers of infrastructure sustains cultural capital of a sort. Identifying, valuing, and contributing to the appreciation of the cultural capital that is the built environment should be an important role for pervasive computing, the latest layer of spatial adaptation. Successful applications toward that goal will become regarded as natural, or at least appropriate technology. Others will just get in the way.

Architecture provides a fixed form for the flows engineered by pervasive computing. As a much older form of technology it has shaped expectations more fundamentally. It remains an important part of our cognitive background. In the relation of environment and technology, buildings are among the oldest, best understood, and least obtrusive of artifices. Quiet architecture may be our most natural technology.

II Technologies

4   Embedded Gear

Universal	Situated
Anytime-anyplace	Responsive place
Mostly portable	Mostly embedded
Ad hoc aggregation	Accumulated aggregation
Context is location	Context is activity
Instead of architecture	Inside of architecture
Fast and far	Slow and closer
Uniform	Adapted

4.1 Universal versus situated computing

When everyday objects boot up and link, more of us need to understand technology well enough to take positions about its design. What are the essential components, and what are the contextual design implications of the components? How do the expectations we have examined influence the design practices we want to adapt? As a point of departure, and with due restraint on the future tense, it is worth looking at technology in itself.

To begin, consider how not all ubiquitous computing is portable computing. As a contrast to the universal mobility that has been the focus of so much attention, note the components of digital systems that are embedded in physical sites (figure 4.1). "Embedded" means enclosed; these chips and software are not considered computers. They are unseen parts of everyday things. "The most profound technologies are those that disappear," was Mark Weiser's often-repeated dictum. "They weave themselves into the fabric of everyday life until they are indistinguishable from it."[1]

Bashing the Desktop

Some general background on the state of the computer industry is necessary first, particularly regarding the all-too-familiar desktop computer. By the beginning of the new century, technologists commonly asserted that a computer interface need not involve a keyboard, mouse, and screen.[2] Ambient, haptic, and environmentally embedded interface elements have become viable, and these require more concern for physical context. But because the desktop graphical user interface was just what made computing accessible to nonspecialists, it remained the only form of human-computer interaction that most people had ever known. By now that model is more than twenty years old. In the time frame of the information industries, that is astonishingly long.[3]

"What's wrong with the PC?" usability advocate Don Norman asked. "Everything. Start with the name. The personal computer is not personal nor is it used to do much computing."[4] Technology experts generally agree that the personal computer was a good idea for its time—twenty years ago. But by now that box is used for so many purposes that it has become unwieldy for any of them. Even its most essential applications—text, graphics, databases, and spreadsheets—have been weighed down under hundreds of commands. A graphical user interface made sense when you could put everything of relevance on the screen, but now there is just too much to see. A single hard disk made sense when its owner could know more or less what was on it, but now it typically contains thousands of files, most of them put there by someone else.

Software piles up. Programmers' impetus to add ever more features has led to bloat. This prolixity is a consequence of logic, which is naturally cumbersome. Each time another option is considered, the number of conditions that must be coded and configured increases. Faster chips run all this code acceptably, but people and organizations seldom benefit.

This lamentable state of the desktop computer results from poor design. As software critics such as Alan Cooper have explained, design for usability or experience too often comes only after this engineering, by which time it is too late.[5] As long as people accept design to be some programmer's notion of what features to add this year, there will always be more features, whether or not they are usable or useful.

And now, as computers move out into the physical world, better design becomes essential. Pervasive computing has been hailed as an escape from the desktop and a chance to start over. On the other hand, unless design can intervene, it is also a chance for computer technology to become even worse, and far less escapable.

For any new approach to design to break out of this feature-accumulation cycle, information technology must change fundamentally, that is, at a level much more basic than a better desktop interface. In essence we face limits to how much we care to do or will consider doing with any one device in one place. More subtly, we also face limits to how much a device can do without better information about its context.

In response, the computer industry now researches aggregations of smaller, more specialized, more localized systems. As Sun Microsystems chief scientist Bill Joy has argued, we are actually facing the evolution of operating systems. The next stage of evolution takes the load off a technology now two paradigms old. Thing-centered computing is coming to be for the 2000s what network-centered computing was to the 1990s and personal computing was to the 1980s. A personal computer was (and still mostly is) based on its hard disk. The disk operating system (DOS) on which this was based assumed that everything one needed was stored locally. A network of computers, on the other hand, was (and presumably will remain) based on packet-switched communications. The transmission protocol (TCP/IP) on which this was based assumed that everything one needed could be made universally accessible on the Internet. This arrangement did not do away with the local hard disk, but it moved information on and off it in a way that was much more significant. Whereas the isolated desktop system encouraged only more automation of tasks, network computing allowed full-scale organizational change. So many possibilities had to be accounted for, however, that the old disk operating systems grew into baggy monsters.

This evolution intrinsically embraces context. DOS assumed everything was just local (but not at all networked). TCP/IP assumed connectivity was universal. What sort of standard will emerge on the assumption that what you need, and with whom you wish to be connected at the moment, is based on where you are? Bill Joy called this a question of etiquette. If our growing constellations of devices and gadgets are to become any less obnoxious than the desktop computers they are intended to replace, they will have to acquire some situational protocols.

This view often finds analogies in electrification. One usual cliché describes the refrigerator: You probably don't think of this machine as an application of an electric motor, and you don't need to know about electric motors to use it; you just treat it as a cold, dark place to store your food. Similarly "information appliances" let you carry out particular activities without having to be aware of any computers that may be involved. As Don Norman urged, "The primary motivation behind the information appliance is clear: simplicity. Design the tool to fit the task so well that it becomes part of the task, feeling like a natural extension of the work, a natural extension of the person."[6]

Norman's calls for "invisible" computers reflect the need for less attention-consuming technology. This echoes Mark Weiser's vision as well. "Invisibility has become the main issue now," he remarked in the early 1990s. "We have been very good at putting computers into the environment, but we have been very bad at getting them out of the way."[7]

The goal of natural interaction drives this movement toward pervasive computing and embedded systems (figure 4.2). There has been something unnatural about how desktop computers ignore many possible dimensions of skill and yet monopolize our attention. In a natural interaction, context would include information that does not require our attention except when necessary. Ambient interfaces would let us monitor potentially relevant information. Haptic and tangible interfaces would use latent intuitive physics.

Yet there is a lot more to embedded systems than there is to household electric appliances. This is basically a consequence of how they tend to link up. Ad hoc physical aggregations of digital devices become systems in themselves. Interoperability becomes critical here. "A distinguishing feature of information appliances is the ability to share information among themselves," Norman noted. "Smart devices are all about internetworking. Information isn't put into your computer so much as your computer is put into a world of information."[8]

You do not have to be a cultural anthropologist to suspect that "anytime-anyplace" universality is not the ultimate technological outcome. For technological reasons alone, we can see that fixed and specialized contexts will accumulate technology and diversify interac-

	Virtual	*Physical*
Foreground	Graphical user interface (GUI)	Haptic interface
Background	Ambient interface	Inhabitable interface (smart space)

4.2 Beyond graphical interfaces

tivity.[9] What is more, we are beginning to see how those contexts accumulate digital technology. Quite the opposite of a global brain (or Big Brother), arbitrary local aggregations of self-connecting systems can become islands of coherence in the chaos raised by pervasive computing. How to achieve this becomes more than ever a question of design.

Many properties of ad hoc networking interest the designers of physical space. When connectivity lets a device discover what other appliances and services are on hand, then design necessarily involves processes that are best understood as local. The sheer volume of technological possibilities alone dictates that activity must be managed by context. Local properties of scale, discovery, protocol, configuration, and tuning all become essential. In comparison to current universal networks, local systems do not require such high-level models in software, and they are less subject to monitoring by external parties.

This emphasis on the local actually points away from Orwellian prospects. Alex Pentland, long a pioneering researcher in smart environments, has explained the advantages of locality by distinguishing "perceptual intelligence" from "ubiquity." "The goal of the perceptual intelligence approach is not to create computers with the logical powers envisioned in most AI research, or to have computers that are ubiquitous and networked, because most of the tasks we want performed do not seem to require complex reasoning or a god's-eye view of the situation. One can imagine, for instance, a well-trained dog controlling most of the functions we envision for future smart environments."[10]

Instead the intentionality must lie in human organizations, abilities, and practices. As we have seen, foundations in embodiment and activity theory help tilt the field of computer-human interaction (CHI) back toward understanding how people play situations. This new era in information technology is not about predictable outcomes so much as richer experience. It pursues much more human-centered goals in natural interaction.

We have seen how embodiment shapes expectations. From a philosophical standpoint, activity in context has supplanted tasks on isolated machines as the focus of information technology design. From architecture we have identified a latent need to map our embodiment onto the world. This pertains to the present discussion in that we feel a deep need to maintain technological constructs whose dimensions resemble those of the human body in architectural space.

Now let us apply this background to recent developments in technology. Physical devices establish possibilities for interaction beyond the desktop. Local models are necessary abstractions for technology-extensible places. Social situations provide design precedents and problems from which to build types. All of this points toward new forms of context-centered design. Whether the various goals are met, and whether this next layer of technology is to become a bane or a boon depends mostly on how many ways of environmental knowing can be brought to interaction design.

Understanding the Components

As computing moves beyond the desktop, what are the essential building blocks that nonspecialists may want to understand? As a way of reviewing recent developments, consider some essential categories in embedded computing technology. The elements and applications of ubiquitous computing are fairly well established within technical circles by now. Different emphases on their particulars distinguish research areas. Among those elements held in common, some should survive rapid changes in circumstance. Already these developments have consequence for the shift from virtual world building to physical computing.

Much of this work begins at the level of demonstrating technical possibility. Concerns for usefulness, organization, and social appropriateness inevitably follow. More so than engineered features, these situational factors cause some technologies to linger at the margin while others explode onto the global scene. It is notoriously difficult to predict how quickly developments will come on the market, and for what initial purposes. So put aside the technofutures for a moment. Here without speculation on their implications is a set as ten essential functions from which pervasive computing systems have been composed.[11]

1. Sites and devices are embedded with microprocessors.

It all starts with the embedded microprocessor. All sorts of things have a chip in them. The pocket radio of the 1960s, that first truly widespread instance of portable consumer electronics, was named for its embedded processor, which if not yet truly a chip, was at least "solid state." A beachgoer could listen to his or her "transistor."

By the year 2000, a mobile phone could pack more processing capacity and memory than a campus mainframe did in the 1960s. A web server could fit in a pocket. An ordinary chip could hold an operating system, a network interface, an Internet protocol stack, and a web client. In its extreme form, a web-capable device could be smaller and lighter than a nickel.[12] In 2001, news stories popularized knowledge of the "smart dust" developed at the University of California, Berkeley. Using solar power and optical transceivers, researchers overcame the usual size constraints presented by powered communications circuitry. At a 7-mm length, the prototypes were still too large to be blown about like actual dust, but for something able to communicate across the bay from San Francisco to Berkeley, these sensor devices were considered small.[13]

Technofuturists jumped on this miniaturization bandwagon with predictions of practical bionanotechnology within a decade. Nanotechnology takes embedded systems to the practical limits of smallness, with microns-wide devices that we would have difficulty understanding as chips. Biotechnology aims to integrate these with living systems. For present applications to architecture, portable gear, and temporarily worn devices, centimeter-sized devices are small enough.

The move beyond the desktop computer is well under way. According to U.S. government research statistics for the year 2000, shipments of new microcontrollers outnumbered those of new computational microprocessors by a factor of almost 50.[14] According to Intel, already more than 95 percent of devices containing microchips do not present themselves to their users as computers.

As demonstrated by surging enrollment in conferences such as those for embedded systems, device engineers have increasingly devoted their efforts to the interoperability of smaller chip-based devices in physical settings.[15] This technical interest occurs because the embedded device is engineered "low." Practical economies of engineering do not always warrant providing a full-service network operating system; devices can communicate at lower levels without that kind of overhead. Such internetworking is indeed vital. Without it, devices must be hard programmed for a particular purpose, like gadgets. With connectivity, however, embedded systems can communicate their status and receive ongoing instructions to and from their surroundings. In contrast to anytime-anyplace universality, this alternative is intermittent and local.

2. Sensors detect action.

If technologies are to keep out of the way, they need to see us coming. If computationally embedded environments are to be useful yet unobtrusive, they have to recognize what is happening in them. Next to the microprocessor itself, the next most basic component of embedded computing is the sensor. To the forecaster Paul Saffo, for instance, sensors have become the "key enabling technology" for computing. Microprocessors themselves led change in the 1980s, and laser optics (storage-rich compact disks and bandwidth-rich optical fibers) were key in the 1990s, "But we are beginning to see diminishing returns from merely adding more bandwidth to our access-oriented world. Now change is being driven by sensors—cheap, ubiquitous, high-performance sensors, or MEMS—and they will shape the coming decade."[16]

Like processors and networking before them, sensors have now reached the steepest part of the cost-reduction curve. Today a programmable network of wireless embedded sensors often costs less than one hard-wired dedicated single sensor circuit did not so long ago. Thus many more applications become normal. For example, a house may be equipped not only with smoke detectors, but also with inexpensive monitors for specific gases such as radon or carbon monoxide. On the streets of London, smog monitors mounted on lampposts transmit data to a nearby server. Devices under development there cost less than 1 percent as much as their predecessors.[17]

This invites a reconsideration of some basic engineering concepts. A sensor responds to a change in state. The medium in which the state exists might be mechanical, electrical, magnetic, hydrostatic, flowing, chemical, luminous, or logical. The change might be a discrete event, the gradual attainment of some threshold, or the establishment of a pattern. In effect, even the simplest mechanical sensor intrinsically serves a logic device, which simply reports whether a change has occurred. For example, a crude trip wire has two outputs: no, no one has walked by yet; or yes, someone just has.

Typically in the history of engineering, some aspect of the physical configuration of a device, relative to its monitored medium, had to contain implicit information about what constituted a meaningful change. For example, a float valve responded to water reaching a prescribed level. Such a device would normally have to be calibrated to fit its setting, usually by means of some dimensional adjustment to a scale. Typically this kind of sensor would serve a single purpose, for which a sensing system would in effect be hard programmed as a mechanical computer. Optical and electronic systems have been more powerful, and many of these have long involved the use of microprocessors, but still they too have been operated in isolation and for a single purpose for which they have to be configured in advance.

Embedded computation changes that. Adaptable programmability is the key to the relationship between sensors and embedded microprocessors. Now the signal from a sensor can be interpreted statistically, over time, and in comparison with background conditions held in memory. This becomes especially powerful in large arrays. Systems for interpreting signals from a field of linked sensors have dramatically increased the capacity to recognize patterns. Wireless networking improves the practicality of distributed fields of sensors. Until recently, any mechanical-electrical sensor has required a full-time, direct, hardwired connection to its controller. Usually that has required a dedicated network, and at best it still had to share traffic on a full-service local area network. Often this connection has been more expensive than the sensor itself. Cost has limited the number and distribution of sensors, which in turn has limited the adaptability of whole systems. Now a field of wirelessly interlinked sensors can become practical. Interlinking can implement much more economical protocols for downtime, for passing or "hopping" messages directly among themselves, and even for reassigning tasks to one another. By analogy, such a field becomes more like a population, like birds that quickly pass the word when a cat is walking around beneath them. Continuous sensor fields can ask: "What trends are developing here, besides discrete events?"

The case of microelectromechanical systems circuits has been especially telling. According to estimates by an industry consortium, the MEMS market is expected to grow from $2–5 billion in 2000 to $8–15 billion in 2004. There were 1.5 MEMS devices per person in the United States in 2000, and that number was expected to grow at a compound rate of over 40 percent a year, to 5 such devices per person in the year 2004.[18]

The accelerometer, one common MEMS application, advances the cause of haptic interface design considerably. Instead of pushbutton clicks, interfaces employing a change of pace in continuous motion invite more skillful, embodied, and unobtrusive operations. The wave of a wand becomes much more of a reality. For example, one Nintendo GameBoy uses MEMS to allow video play by tilting rather than clicking. At larger scale, accelerometer-based gyroscopes replace spinning mass systems for navigating high-speed Acela and TGV trains.[19]

Also relevant to embodied interaction design, pressure sensing has become more practical. A pressure-sensitive resistor now costs little enough that it can be used in casual applications. When distributed in an array, these can make a building surface responsive to human presence (figure 4.3). In a demonstration of this possibility, Hewlett-Packard's David Cliff has rigged a pressure-sensitive dance floor

4.3 A responsive building surface: Aegis Hyposurface. (*Courtesy of Mark Goulthorpe, dECOi.*)

through which the activity of dancers is fed back into a musical selection or real-time composition. If it were coupled with wearable biofeedback devices such as heart-rate monitors, or motion accelerometers, and with breeder algorithms in the composition software, this could provide a fresh trance scene without recourse to chemicals.[20]

Although the range of physical possibilities expands, nevertheless most sensing tends to remain visual. Computer vision has been intrinsic to many agendas in pervasive computing. This is because identifying the actors is such a fundamental issue in context-aware software and because tagging is not always welcome, practical, or sufficient. Abilities in vision are relative, of course. An assemblyline robotic arm can "see" the objects it needs to pick up, but nothing more. Almost as quickly as the state of sensing advances, however, old fears about panoptic social control reawaken. Cameras appear in more places all the time, and we are rightly afraid of them.

According to ctrl [SPACE], an exhibition and publication from art and media center Zentrum für Kunst und Medientechnologie (ZKM) in Karlsruhe, Germany, it is not only visual sensing by cameras, but tracking by various forms of what the organizers called "data-veillance" that should concern us.

Sensors are not only the most essential but also the most unnerving of the new technology elements. Concerns about privacy make most of us unreceptive to this technology. But that is a larger social question. As Scott McNealy, chief executive officer of Sun Microsystems, said in 1999: "You have zero privacy anyway—get over it."[21] We can only hope that with billions of sensors in place, there is too much information for any software, much less some government agency that cannot even master its own internal communications, to interpret successfully.

3. Communication links form ad hoc networks of devices.

Local-hop sensor fields are just one example of how networks may temporarily form. Instead of being intensively planned and rigged, pervasive computing depends on unplanned communication. To avoid technological overload and privacy losses, connections are opened only where necessary.

Practical links can be fixed or portable, specialized or general, constant or intermittent, and passive or interactive. Their usability depends on interrelationships as in an ecology. Patterns of unplanned communication may allow unanticipated local capacities to emerge.

Even among purely nomadic technologies, context has begun to play a role. The radio-based Bluetooth protocol, introduced in the late 1990s, quickly became a standard for linking nearby devices. Five years later, a universal serial bus (USB) chip could be used to make peer-to-peer connections between devices formerly requiring a host computer such as a desktop PC, and at a bandwidth considerably higher than wireless Bluetooth. Not all linked objects will benefit from a full-featured web browser, of course. More will run some slimmer set of communications.

Thus when current developments start from the idea of a web connection, they often work toward much "lighter" (lower communications overhead) and more flexible connections. Web protocols as we know them do provide a convenient way to use interactive computers as we know them. They let us monitor embedded computers and sensors in applications where the networking cost is justified. Server-side computations can employ powerful computers and high-level programming languages to manage smaller devices that are not so easily codable in themselves. Meanwhile, some web servers can now fit in one's hand—or in a hand-sized object.[22]

But smaller objects are codable—that is the whole point of the hype. As has been fairly common for some years now, the Java model lets programs developed at a high level be executed locally on small machines. Most microprocessors have the power to serve as a Java "virtual machine"; such processing logic is not the main engineering constraint (power supply, for instance, can be more the issue). Moreover, devices running Java software can poll their vicinity to find out what other relevant devices are available for interconnection. In 2000, Sun Microsystems' JINI standard introduced a "network dial tone" by which such polling could occur economically. The increased connectivity that results from such a discovery service increases the likelihood of a division of labor. For example, a small Java device that needs some interactivity but cannot justify buttons or a display can offload its interface to a larger device in the vicinity. A larger device monitoring the activity of several smaller ones can invoke their local computations remotely. Furthermore it can download different software onto those devices to change their role. Networked communication thus dramatically increases the capacity of a local collection of devices to adapt to incidental conditions.[23]

Ad hoc communication thus adds new capacity for the appropriation of context. Fixed resources may suggest, and to some extent become adapted by, different configurations of software. This is fundamentally different from an anytime-anyplace uniformity, and it is an important property of place. Adaptive reassignment of locally networked devices raises the prospect that pervasive computing constitutes a new type of recyclable resource. Dan Siewiorek, director of the Human Computer Interaction Institute at Carnegie Mellon, has called this "renewable" computing. This occurs not only at the level of local networks, but a larger and smaller scales as well. "The micro level considers harvesting energy from the surroundings. The infrastructure level addresses renewable structures and how services emerge as the mix of devices evolves. The macro level considers the transformation of information sharing and impact on existing systems such as transportation and industry."[24]

Renewable, location-based infrastructures constitute quite a different future from that of disembodied cyberspace; yet the present state of pervasive computing is neither of those. Communication between interactive portable devices dominates current developments. These are still in the foreground; they are not yet much of a periphery.

As with any technofuture, there is a big difference between the newly possible and the generally practical. Intel director of research David Tennenhouse has cautioned: "Radical innovation will be required to bring networking costs in line with the $1-per-device price structure of the embedded computing market."[25]

4. Tags identify actors.

Contextual awareness begins from an ability to recognize who or what is present. Pattern-recognizing sensors can detect some of this, but for habitual applications, or for ambiguous conditions, recognition becomes much simpler when something is simply tagged. This is forecast to grow into a billion dollar industry in America in this decade. A radiofrequency identification tag (RFID) cost less than a dollar in 2001, and is expected to cost less than a penny with a few years.

Tagging based on earlier information technologies is already widespread—especially the bar code. As a great many news stories proclaimed, 1999 marked the quarter century anniversary of the uniform product code (UPC), which is the standard that brought widespread success to optical tagging, first of groceries, (although the first items imprinted with bar codes were railroad cars), then with most retail goods, and sometimes even people themselves, such as hospital patients. The electronic product code (EPC) that has been designed to replace the UPC takes this further by giving a unique number not only to each product, but to each unit of that product.

Tagging has abundant architectural precedents. Ornament, inscriptions, and signage have all in effect tagged the built environment with information intended for particular purposes. In some cases, especially where automobiles dominate traffic, signage has become more prominent (and often more invested with design) than the buildings themselves. Tagging now complements the signs and symbols that architectural theorists know so well with more diverse forms and smaller scales than has been practical before. Some of these are quite pervasive, such as tagging places by means of their coordinates in a global positioning system (GPS).

The addition of computation to tagging yields a lot more applications besides pricing and gatekeeping. Just about any barcode application is coupled to an inventory control system. Smart touch-sensitive counters, such as those that have been demonstrated on bins in parts warehouses, can conveniently monitor the flow of goods outside all the usual point-of-sale documentation.[26] Any place that large numbers of such tags are read, particularly the retail point of sale, in effect becomes an important component in the use of space. The overall flow in a system of tagged items might only be visualized by analysts at some central hub. Major retailers such as Walmart have thus been among the first to pilot use of RFID tags on a large scale for tracking low-cost merchandise as it moves off the "smart shelves."

A tag can be used to summon annotation as well. For example, Sony has demonstrated a system for summoning biographical and professional data on office occupants by means of tags from bar codes on office doors (the Navicam).

Tagged items can function as physical tokens in hybrid physical-digital systems. In essence, a token represents an abstract arrangement with a physical symbol. This is not to say that we want to embed a text of vows into a wedding ring; tagged tokens seem to take the more mundane form of yellow stickies and refrigerator magnets.

For example, in "Triangles," one of the first broadly recognized projects in the Things that Think consortium at the MIT Media Laboratory, palm-sized physical triangles served as tags for project data; attaching them to one another or waving them in front of a camera computer triggered operations on a desktop computer. Early on,

Hiroshi Ishii's group demonstrated the principle of using tagged objects as "phicons" (physical icons) or "metablocks" as interface widgets.[27] One obvious connection is to link physical tokens to websites. Even a banana at the supermarket has a web address printed on its price code sticker.

From a technological standpoint, smart tagging brings software into the physical environment by means of small, affordable Java virtual machines. For example, "iButton" is a 16-mm portable Java platform that can be affixed to a keychain, a ring, a wall, or the side of a computer. As of 2002, more than 65 million iButtons were in circulation, and one could order them in lots of ten thousands on the Internet.[28] As buttons, one could sew these onto a shirt, or make smart jewelry from them. As an inescapable part of embodied computation, computation moves onto the body.

Tagging *people* raises a lot more questions. We all carry identification cards with magnetic stripes. A Harvard faculty ID will cease to let you into your old building less than a month after your appointment has ended. In early 2002, Hong Kong introduced smart ID cards that included biometric information such as a thumbprint, as well as a picture, birth date, and address.[29] However, a card is still vaguely private in that you keep it in your wallet.

"Biometric" identification has been made topical by recent interest in domestic security. It involves no stealable passwords or tokens. One of the first "biometric" authentication devices to come to market at an affordable price ($200) was offered by Ethentica.com. In 2001 the company began offering digital fingerprint authentication; owners could attach fingerprint security to individual data and applications, as a substitute for remembering and typing.

Tagging becomes more blatant wherever people agree to wear badges. In a world full of mobile workers, badges appear to have become an accepted fact of life, at least in workplaces. Social scenes have their own subtle badges of standing too, of course, and when these start polling and linking it can make for quite a scene indeed.

A badge that functions within a limited radius, that allows passage of a particular perimeter, and which (occasionally) helps customize local resources for its holder has not surprisingly been among the earliest testbeds for pervasive computing. A smart badge brings all the technical, and more important, all the social issues into focus. Badges were intrinsic to Xerox PARC's early work on pervasive computing. These were inherently contextual as well; they functioned in specified locations only, in confederation with other digital devices at larger scales. Similarly, the Olivetti Smart Badge, often cited as an earliest precedent in this technology research, was a site-based, context-aware application.

5. Actuators close the loop.

In a sense, whenever a system regulates itself by monitoring its own performance (i.e., with a feedback loop), some rudimentary intelligence is implied. By this reasoning, even an ancient water clock was "smart." Similarly, a household thermostat, based on thermal expansion of a copper coil to meet an adjustable electrical contact, is an analog computer of a sort. As a switch it is an actuator—a device that alters a system's state when it is triggered by appropriate conditions.

Modern industrial process engineering has been based on devices, such as the servomotor, that translate electronic signals into physical actions. Digital systems expand the linkage and logic behind these control signals. Besides sensors, then, embedded gear advances and diversifies the role of actuators.

Technofutures have been rife with these. Wherever industrial processes and products have employed feedback control systems, popular imagination has extended their application to domestic, social, and recreational uses. Today the old Jetsons fantasy about mechanical devices that pamper us (or the much more mod-cybernetic version of same from the magazine *Archigram*) has been renewed with digital devices that adapt to us. Neo-Jetson enthusiasm for the smart appliance revives all the questions of how to keep actuators in the periphery.

We want environmental systems to keep out of the way, but we also want them to do something. We might be interested to see "environmental controls" do something more interesting than just stabilize indoor climate. This is not all fantasy; many of these exist. For half a century, Disney imagineers have excelled at putting actuators into small objects, for example. An "animatronic" theme park character contains dozens of servomotors. A recent luxury car is also loaded with actuators, and many of them are coupled to fairly sophisticated computers. Instant airbags and antilock brakes receive most of the press, but fuel systems, valve systems, steering systems, vibration isolation, suspension, even seat adjustments employ meters, timers, gates, and especially a lot of fast, small actuators to improve whole-system response.[30]

In architecture, applications began with resource management systems, particularly those for energy, which especially in America have seemed anything but smart. Computer-driven actuators can adjust sunshades, schedule peak demand loads, manage the much more complex states that result from much more localized control, and so on. Maintenance systems have spread as well; sensors built into structures can identify deterioration and signal for upkeep before failures occur. Bridges and dams, whose failure would be catastrophic, increasingly use this technology.

Building geometry itself has sometimes seemed suitable for kinetic systems. Embedded kinetic elements manipulate other components of buildings, like tendons moving skeletons. Deployable kinetic elements are temporary structures, like tents. Dynamic elements are independently mobile, but are affixed to buildings, like doors and canopies.[31] For example, responsive systems have received some press in Enrique Norten's design for the Educare school gymnasium in Guadalajara, Mexico, in which sensor-controlled enameled steel panels open and close, in the words of the architect, "like the scales of a fish, or the feathers of a bird."[32] In the prominent first American building by Santiago Calatrava, a pavilion for the Milwaukee Art Museum, the 200-foot sunshade over the reception area opens "like the wings of a bird."[33] These sculptural applications have moved sensing into the profession's imagination in a way that other climate-control systems cannot. And by monitoring atmospheric rather than social conditions, they have done so in a way that is not alarming.

The physical environment abounds with opportunities for improving commodity, firmness, and delight through the application of intelligent feedback systems. Commodity is largely a matter of life-cycle economy. Energy, materials, and space are wasted at prodigious rates in built environments.[34] Firmness invites applications for systems that provide safety and security in a more intelligent, less obtrusive way than today's building codes. Meanwhile, delight is up for grabs. Places that afford some participatory adjustment on daily and seasonal cycles have long been alleged to be more satisfying than those that are uniform. Spaces that subtly reconfigure themselves according to their occupants and use can cause paranoia or delight, depending on how intelligently they are designed.

6. Controls make it participatory.
If all this technology were completely automatic and able to function completely passively, it would be out of the way all right—and more frightening than ever. Instead, smart systems need to be operable where it is appropriate. It might be preferable to configure some systems just once, to adjust others occasionally, and to incorporate a few into daily routines. Like a remote control for television, some might be twiddled for amusement. Like a musical instrument, some smart systems can facilitate personal growth through skilled practice.[35] These are basic principles of interaction design: Know when to eliminate an obsolete "legacy" operation, when to automate, and when to assist an action. Know how to empower, not overwhelm.

"Pushbutton" convenience was a hallmark of a modern age only recently freed from bodily work. The industrial button was once understood as an abstraction. The automated operations it triggered were still fresh substitutes for something much more tedious. The postindustrial button frequently lacks a referent in bodily experience, however. This is partly because its logic leads to cumbersome convolution. Hundreds of buttons, or hundreds of expressions for entry by buttons, depart from the realm of comprehensible experience. This is why the household videocassette recorder has become a standard emblem of incomprehensibility.

Haptic interface strategies based on gestures, gliders, and motion sensing provide alternatives to the current excess of clickable buttons.[36] For example, the tilt-and roll technologies of the MEMS-equipped Gameboy can be carried over to the a handheld computer.

This may be used for gaming at first, but once people are comfortable with the operation, it can be used in other applications. Motion Sense plug-ins extended capacity this to Palm devices in 2001, for example.

Where the sensitivity of applications warrants investment in much more specialized interfaces, haptics technology has advanced considerably. For example, the ReachIn interface, used in training persons for medical procedures, combines the Phantom pointing device with a reflected transparent screen that puts a virtual display right over the active hand. With a repertoire of specialty software for moving and modifying surfaces, and for rendering texture and friction, this arrangement has delivered the best multimodal haptic interface to date. "Haptic rendering," the process by which virtual surface properties are communicated through an ultrasensitive force-feedback device in real time, bears many analogies to the graphics rendering of a decade earlier. Like graphics, these algorithms eventually end up in hardware, on a chip, where they become practical for more casual application, such as in everyday geometric modeling in computer-aided design.[37]

We know from traditional craft that finesse requires some touch but not necessarily full embodiment. This is an important distinction. The usual objection has been that symbolic processing requires us to sit still. Relative to traditional work practices, this was more or less true. But as we have discovered from the first few decades of creative computing, the active participation so conducive to learning and expertise wants more than a mouse and a screen. Even within the crude window-icon-menu-pointer (WIMP) technology, there is enough affordance to support talent. The way is wide open toward physical interaction design: active controls are integrated with comprehensible, satisfying things.

7. Display spreads out.

For a precedent in ubiquitous information technology, Mark Weiser would point out text. Text really is ubiquitous—you are rarely out of sight of several pieces of it. Take cars; as anyone who has time to study them amid gridlock on the freeway knows, the back end of a car can signal you with at least five text sites: the bumper sticker, the license plate frame, the characters of the plate itself, the dangling sign cautioning about whomever is on board, and the window decal.[38] The back of a cereal box is a cacophony of texts and images vying for the awakening person's momentary attention. And, in an act that would most likely astonish a visitor from any other century, millions of people think nothing of going out wearing unpaid advertising in the form of logos and messages on their clothes.

This is a tale worth its many recitations. Once upon a time text was scarce, and was generally confined to the library. Occasionally some of it would be chiseled onto buildings, but that too was a rarefied setting. Modern printing obviously changed all that. "This will kill that," said Victor Hugo of the printed word's advantage over the inscribed building. In the past 50 years, printing and photo composition have moved text and images into unprecedented contexts. Next, text displays came alive. The word processor is one reason why most of us tolerate computers. More recently the scale of displays has expanded down to handheld devices and up to cover whole walls. As evidence that information truly has become ubiquitous, the text screen on a gas pump scrolls an advertisement while you are filling your tank.

The idea that how we do things with symbols depends on their scale and position relative to the body is fundamental to pervasive computing. Something you read inclined on the sofa should be a different size than something you read as you step out of sidewalk traffic into a doorway. At 10 feet high an image reads very differently than the same content at 5 inches. Scale was intrinsic to the original notion of ubiquitous computing: Weiser's group at Xerox PARC made a typological distinction among tabs, pads, and boards.

Lately it has become possible to move text between many scales and surfaces. Researchers from Sony have demonstrated ways to drag an image off a laptop computer and onto a wall.[39] The greater the variety of display surfaces on hand, the more appealing such mobility becomes. For an example of free-form display, IBM's Everywhere Displays project combined projection with detection on an arbitrary surface such as a tabletop or a wall.[40] In effect, this coupling turns the surface into a crude wireless touch screen.

The expression *augmented reality* has generally been reserved for conditions in which a virtual display is overlaid onto a physical scene. Sometimes this occurs on the eyeglasses of a viewer. Such applications have been pioneered on assembly lines and in equipment repair, for example, where annotation of viewed objects is needed but the hands have to be kept free.

Nevertheless it is the live display in the fixed setting that may most characterize embedded systems, and that brings pervasive computing into architecture. Again the earliest examples were in the laboratories. At MIT, Alex Pentland's ALIVE project team maintained a wall-size screen suitable for animated characters.[41] Hiroshi Ishii's "ambient ceiling" set an early standard of being in the periphery.[42] It was based on projected images of ripples like those on the surface of a pond, whose frequency and intensity were mapped to local measures of ambient conditions such as network traffic or the number of people on site. Wall-sized displays have been important to more recent interactivity research at Stanford University. These "murals" provide much more substantial content than previous projected displays, while keeping the control elements to a minimum. Laboratory director Terry Winograd has often used architecture metaphors to describe how, in the design of interactive experience, the whole is greater than the sum of the individual technical components.[43]

Many people think of large-format displays in terms of the year the NASDAQ sign went up. This outdoor installation demonstrated that besides size, another factor important to ubiquitous display is its ruggedness. A display that can be left out in the rain opens a very different realm of imagination.

Next, the literal ground itself becomes interactive. With the spread of positioning systems, which in effect make anyone who carries such a system into a live cursor, the city plan itself becomes a living surface.

Perhaps because we have been culturally conditioned to fantasies being visual, or because the dominant cultural media of the last century would have been wild fantasies in the one before that, prospects in display technology make it more difficult to avoid technofuturism.

8. Fixed locations track mobile positions.

Positioning technology has exploded. In a 2002 interview, GPS pioneer Per Enge observed that in contrast to original projections for an eventual market for 40,000 GPS units (mostly in military applications), at this point 100,000 were being produced each month.[44]

"Let's put GPS in necklaces and dog collars. Everything that moves should have GPS," wrote Kanwar Chadha, chief executive officer of SiRF, the leading GPS chipset manufacturer.[45] "This kind of stuff has enormous potential for abuse by the authorities, or by anyone who can break into the information," wrote Emily Whitfield, a spokesperson for the American Civil Liberties Union.[46] Concerns include not only the usual fears of governments monitoring their citizens, but also criminals tracking their prey.

Among the ways to achieve more natural interaction design, there is none quite so obvious as using position for input. You do not have to be a genius to understand the potential of a device whose sole instructions are to take it somewhere and turn it on. Maybe a third step would be to pop in a filter for what you want to know: plant identification, history of a city, bar hopping, tracking your friends, finding your tribe.[47]

Because space is such a fundamental category, position can serve as an index to a range of information services. Any body of data that can be "geocoded," that is, assigned a position as a key to a record in a relational database, can be delivered intelligently through geographic information systems (GIS).[48] With mobile communications, the information can be delivered where needed. A GPS completes the loop; position data are used to query huge spatial databases that report relevant information back to the position being described. That information can be highly thematic, and not just the stuff that usually shows up on printed roadmaps. For example one might use a mobile GPS-GIS system to study vegetation patterns, ethnic neighborhood boundaries, or current nightclub scenes.

When coupled with tagging, positioning technology helps track things that move around. This helps answer such fundamental questions as "Who is here?" and "What are they doing?" Like the products in a retail supply chain, elements of other networked distributions become documented and their flows become better modeled. Knowing where things are leads to natural economies of routing. Even a sidewalk vendor knows there is value added in this. Tracking also improves the most rudimentary aspects of sport, as any hunter knows. Tracking distributed fields of movements raises new prospects in scientific and recreational visualization. Social delights and abuses quickly occur to the technofuturist imagination. When it is combined with architecture's role of arranging bodies in space, for example, tracking seems more like a security application than play.

Positioning systems also cater to the geographically unskilled. In the late 1990s, Hertz began offering "NeverLost" service automobile navigation system in certain rental areas. "Whereify," a wristband device for tracking a child, was one of the first commercial GPS wearables (figure 4.4). This was just the tip of the iceberg on intelligent transportation systems (ITS), which in turn were just one sector of what was already a multibillion dollar market in geodata.

4.4 Wherify GPS Locator, a location-tracking wearable for children. (*Courtesy of wherify.com.*)

With the cost of a GPS chipset falling through $10, and with the repeal of its resolution reduction for nonmilitary applications as of April 2000, the way for computers to annotate the physical world was been cast wide open. "Urban markup language" does not yet exist.[49] However, in 2001, a geography markup language standard was introduced.[50]

9. Software models situations.

As hardware becomes less expensive, more diverse, and more plentiful, software becomes more challenging. Representing scenes and situations becomes the essential challenge. Knowledge representation remains perhaps the fundamental challenge in software. As evidence for this, the discourse has shifted from artificial intelligence to ontology; that is, to representing the existence of people, actors, and things and their contexts. "Who is here and what are they doing?" recalls Laurel's foundational notion that the purpose of computers is to let people take part in shared representations of action.[51] Add the many components of pervasive computing to the means, and the pursuit of these ends becomes more interesting. For example, a system may begin to model a physically proximate area by polling local ad hoc links between known mobile tags and devices. Protocols, etiquettes, and other such aspects of social framing become more essential than they were in earlier, desktop realms of computing. Architectural frames, and—despite exaggerated reports of its death—environmental scale and geometry help organize so much information.

Location models appear so critical to the problems of implementing contextual awareness that they deserve much more discussion in this review. Chapter 5 explores the relations between geometry, the geodata industry, and sensor-actuator systems, all in relation to the problem of representing actors on stages. Chapter 6 assembles one possible set of situational types for which such models may serve. Persistent goals in knowledge representation carry over from the world of ambitious desktop artificial intelligence to less versatile, but more numerous instances of information appliances, smart spaces, and interactive environmental controls. For example, the metaphor of the butler still obtains. Like a butler or personal secretary, some contextual computing must anticipate some of our needs before we do, and must carry out some of our business without our needing to know about it; but it must maintain our identity, accessibility, and etiquette where appropriate in the processes.

10. Tuning overcomes rigidity.

Even before any software location models are implemented, even the crudest aggregation of hardware and links must be lived in, lived with, and tuned. Much of the place-centered character of situated interaction design comes from the fact that any fixed collection of devices has to be integrated. Tuning consists of incremental adaptations of configurations and settings based on a qualitative, top level interpretation of the performance, and in best cases, the "feel" of the aggregate. Acousticians tune concert halls. Game developers tune the variables and constraints in strategic simulations. Even when engineers balance complex systems using mathematical models, some tuning creeps in. The prevalence of tuning in today's culture of technology usage is demonstrated by the spread of the word *tweak*.

Tuning includes incremental growth and change. How are new devices added? What model underlies the world in which all of these interoperate? Must the whole system be rebalanced each time it incorporates another element? Steve Shafer from Microsoft Research has argued that not only is tuning necessary, but indeed it becomes one of the central knowledge representation problems of the emerging generation of research.

The most obvious thing about ubiquitous computing or intelligent environments is that they have a lot of devices talking to each other. Accordingly, one of the first questions in building such systems is how to get these things to talk to each other at all. This raises questions of network protocols, distributed object programming systems, etc, and much research proceeds along these lines, as though that were sufficient for ubiquitous computing. But, assuming the connections are made somehow, the deeper question is, how can we make these interactions meaningful? What would it take to call device interactions meaningful?[52]

Then the systems start to tweak themselves. In contrast to a sense of place, consider places with sense. Smart spaces recognize at least something about what is going on in them, and then they respond.

Some of this built-in understanding now can reside in easily adaptable software, some can be implicit in occasionally reconfigurable arrangements of furniturelike hardware, and some remains better off being built in. It is the interrelationship of these that needs design.

The question exists of whether each smart space must be built in an ad hoc manner, or whether standards and reusable software components will emerge. Such an infrastructure would be somewhat analogous to the "event handler" typical of interactive desktop windowing and multimedia scripting systems, but would be infinitely more complicated in the continuous context of a physical space. Multiple actors, ambiguous aggregations of objects, and unreliable data streams from distubted sensor fields all replace the mouse click as inputs to be interpreted. Researchers in this must extrapolate from what they know. Programmers from the AwareHome project at Georgia Institute of Technology have developed "context widgets," for example. These are analogous to desktop widgets, reusable modules with fixed sets of callbacks to other elements of the system.[53]

From a systems engineering standpoint, tuning is a matter of regulating the transfers between component devices to achieve an overall system performance that is not easily predicted by numerical methods. Quantitative analysis seeks optimization, which it tries to predict with indicated solutions to mathematical models. Few design problems afford such determinacy, however. Under- and overconstrained problems produce a complexity where human judgment is better than predictive formulations. Here the approach to tuning, and to the design philosophy it represents, is more propositional. This may sound exactly like the "try it and see" approach to design improvization that decades of numerical analysis have sought to overcome, but more accurately it illustrates the need for a partnership between prediction and invention in creative problem solving.

If the tuning of smart environments were to rely exclusively on ad hoc inventiveness, work would proceed slowly. If it were to reduce its considerations to functions that could be modeled more predictably, this effort would produce sterile results. In between these approaches, something is needed in the way of continuous, if not fully formalized knowledge. Invention needs to play off convention. Unless engineers are to face a debilitating agglomeration of gear, some aspect of context has to help with tuning and protocols.

That aspect is type. Persistent structures of form and environment should be able to accomplish half the work of tuning aggregations of portable and embedded technology. If, for example, one is tuning smart gear for a café, a lot of the work should be accomplished by the fact that this is a café. Location and type have to matter. Otherwise, with everything possible all the time, mostly chaos will result.

5

Location Models

Even at a purely technological level, location still matters. Location models prove essential for pervasive computing. Now those representations include architecture.

This is basically a question of representing action. Who is present, and what are they trying to do? How do they receive cues and perceive protocols? How do they know, and when do they care to know, the state of the systems they have encountered?[1] We have seen how perceptions of the possibilities in an environment establish a foundation for interaction design. How much of that can be modeled, and how much must remain implicit?

Location models are not just maps of physical position, but are also representations of activity and organization. As descriptions of work practices, such models became common enough in management amid the networking boom of the 1990s. These were mostly about the intrinsic relationship of information technology and organizational change. No longer just about procedures and automation, computing shaped the very contexts and communities of knowledge. Technology designs thus depended on how organizations went about representing work.

As long as the desktop remained the stage for information technology, location models seemed almost irrelevant. Indeed for a while, many people seemed willing to take the (metaphorical) representation for the (virtual) reality. However, continued human expectations for embodiment and periphery have turned the tide. As we now take mobile devices out into the physical world, and increasingly bring them into contact with intelligent environments built from embedded systems, our digitally mediated actions truly must take place somewhere.

The representation of contexts now becomes an essential challenge to designers of information technology. Conventionally this has been the work of architects, planners, and the allied disciplines of the physical environment. If architecture and interaction design are to benefit one another as disciplines, they must work together on location models.

To an architect, a model chiefly represents form, but to other disciplines, a model may represent behavior, information flows, or decision sequences. To a software engineer faced with ad hoc networking, several of these approaches appear pertinent. Nevertheless contextual computing begins from the physical geometry. We have seen how this is a question of scale. Without some sort of local model, and without some sort of physical scope for local connectivity, pervasive interactivity quickly becomes too complex. Location models must tame this complexity with representations of presence, protocols, and better-presented possibilities for action.

Geometry Still Matters

By now it is common knowledge that at some level the net "negates geometry."[2] It is all about being able to obtain information without having to know where it is stored. If location matters, it is more likely logical than physical. Factors of access, bandwidth, addressing, and security affect who and where you are on the Internet. Position does not.

But even if "you are your address," as is commonly said, that address tends to be somewhere physical. Internet Protocol (IP) addresses are often bound to the hardware addresses of connection devices, which in turn are managed by physical location. For example, dynamically served addresses are only valid within particular security boundaries, and most of those are limited by physical connections and service areas as well. Server farms are big, power-hungry physical sites in locations convenient to the companies that use them. Fixed infrastructures support most mobile systems. Some of these, such as the global positioning system, track position continuously. Others can infer the position of mobile devices at short range as those communicate with fixed network nodes. Others add their own position to information taken from active transactions. When you swipe a magnetic card to enter a building, you are declaring your location.

All this affects us. How many mobile phone conversations casually begin with a declaration of location? How many monitored physical perimeters does each of us cross in a day? Recent obsessions with security make us think about that. Still more than any government, and a huge economic power by any measure, it is the target marketing industry that wants to know where you are.

Geolocation services on the web can trace the country and most often the city of origin of connections to clients' sites. This is useful for setting language, observing music broadcast rights, restricting sales of banned goods, and of course defining saleable market segments. Whether laws belonging to physical jurisdictions should apply to the Internet remains one of that medium's most significant questions, for if every state could insist on web publishers modifying content to fit local regulations, the net would grind to a halt. On the other hand, publishers often voluntarily modify content on the basis of user research that employs geographic tracing. As companies such as Akamai and Quova took geocoded market research online, the business press declared that while distance no longer matters, location most certainly does.[3]

Meanwhile, according to many early developers of intelligent environments, the existing representations of physical space are not enough. The act of bringing a mobile device into contact with a site-embedded system dictates a need for location models, and for several reasons (figure 5.1).[4]

First, there is no guarantee that casual ad hoc connections will make use of, or have their implicit locations inferred by, fixed infrastructures of the Internet. Much interoperability will be simpler, and at lower networking levels, than that.

Second, many situations involve physical relationships that occur at a finer resolution than can be recognized by a global positioning system, or that occur in urban indoor sites not necessarily in view of that system, or that involve spatial relationships more particular than simple co-location.

Third, specialized site-embedded systems cannot operate independently in large numbers. They need to corroborate their respective representations. Individual systems for sensing motion, establishing position, or controlling environmental elements may accomplish their work without the need for a location model, but the more such systems aggregate and accumulate, the more they need to share some representation of who else and what else is present. Otherwise they may produce contradictory results, as well as a great redundancy of hardware.

Why local technology?

1. Pervasive computing is fundamentally a question of representing action. Who is present, and what are they trying to do?

2. If mobile systems are to become less obtrusive, and more useful, they must reflect encountered protocols.

3. The death of location has been overrated: Net addresses are bound to the hardware addresses of connection devices, which in turn are managed by physical location.

4. Specialized site-embedded systems cannot operate independently in large numbers, but must corroborate one another.

5. The geodata industries are exploding, and geocoded marketing remains the most powerful alternative to that ever-elusive goal of "marketing to one."

6. Architectural elements of physical space often frame and cue actions. Sites of interaction involve geometric relationships at fine resolution.

7. Accumulations of technology generally need to be housed, owned, maintained, and tuned. And those accumulations may still represent owners to their constituents.

5.1 Arguments against universal technology

Fourth, architectural elements of physical space often frame and cue actions. When somebody closes a door, they may wish that act could also stop any incoming phone calls or outgoing webcam feeds, although usually it does not. When a mobile device polls its vicinity for available services and connections, it should be able to filter its findings by spatial accessibility as well. For example, it should rule out a connection that is just a few feet away but on the other side of a wall in another firm's offices.

Fifth, few of us care to declare our locations to passive systems, especially often. Our very presence in one kind of space must serve as consent to take part in its technical environment, but in another space should indicate our desire for anonymity. Right now we swipe cards, key in passwords, and plug in cables to declare our presence. There is a dilemma in this. Which is preferable, being subject to monitoring or having to take active part in identifying oneself? If the latter, does that require carrying and using a device, such as a card or key, or merely assenting to the presence of a recognition system, such as a scanner? These are questions of how to present oneself. Traditionally such questions are answered by social customs.

The only alternative is architectural: as always, physical locations come with different protocols. Architecture instantiates particular intentions, etiquettes, and actions.

Each of these factors points to a general need for spatial modeling of digitally mediated action. In contrast to the usual assumptions about formless dematerialization, the rise of pervasive computing restores an emphasis on geometry. The hyperspatial, geometry-negating aspects of electronic communication do not recede, but increasingly are complemented. In locally intensified islands of smarter space, interactivity becomes a richer experience. Generally it does so with the recognition of relationships in scale. These may be from person to person, person to group, group to site, or from person to building. Generally there is much to understand and build, then. Location models present greater technical challenges than desktop computers, but they also offer more kinds of realizable rewards.

In terms of practical implementation, Microsoft researchers Barry Brumitt and Steve Shafer have presented the most coherent case to

 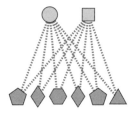 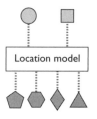

| Ad hoc, direct interconnections | With more devices, interoperation quickly grows complex and redundant | A layer of context software improves capacity and flexibility |

5.2 The necessity of software abstraction

date for beginning to incorporate geometry in pervasive computing. "For computing to move off the desktop and be accepted, it must have a comprehension of physical space which is related to that of the user, else the proliferation of smart devices will only increase the complexity of the user's experience, instead of simplifying it."[5] Their project Easy Living has emphasized the importance of ontology for particular spaces. In order to avoid perceptual inconsistencies and hardware redundancies in smart space systems, a representation of the existence of persons and devices in a space must be constructed as a service independent of any particular application. "Resource extensibility is greatly eased by the abstraction layer that a geometric model provides."[6] In this kind of model, each mobile object and each fixed location has a physical extent of service. Not only can the latter be used to reduce the available connections to a manageable number, but also these extents can be used to reconcile errors and ambiguities among different sensing and positioning systems, and even to discover possible functional relationships (figure 5.2).

Thus the technical argument for geometry builds from first principles of pervasive computing. As we distribute digitally mediated activities beyond the desktop by mobile and physically embedded systems, the relevant relationships between devices can no longer be confined to their network addresses. Ad hoc links between mobile devices and their surroundings usually occur at specific sites or within particular ranges. Physical constraints may determine whether such links are possible or appropriate.

Geodata

Location models belong to a much larger domain of spatial information systems. Led by the explosive growth of geographic positioning systems and geographic information systems, and with extensive applications in security, ecology, planning, and marketing, to name a few, the geodata industries have built up to trillion-dollar volume in a world where geography was supposedly obsolete.

Statistical market research has become such an inherent fact of modern life that we forget how recent a creation it is. Consider a story from the time before the spread of computers allowed the widespread use of such numerical analysis. When in 1962 the United States Postal Service proposed a new kind of delivery aid called the Zone Improvement Plan, civil libertarians and media critics protested that this plan would reduce people to numbers.[7] Today your ZIP code tells more about you than any 1962 postal clerk, civil libertarian, or corporate communications officer could ever have imagined. More than any other single indicator, ZIP correlates with how you vote, what kind of money you earn, which kinds of actors you prefer to see in television commercials, and what kinds of places you might frequent.[8]

Now geodata become available on site. The process of gathering them becomes more widely distributed as well. The processes of sensing, modeling, maintaining, and delivering spatial information all become applications for pervasive computing. For example, the visual flyovers of satellites that have dominated remote sensing in geodata collection are now complemented at a local scale by installed sensor fields. Stereogrammetry and image analysis have been useful for documenting the lay of the land and deducing the presence of certain features in an area; they are the legacy of military reconnaissance. Installed remote sensing allows other uses: counting traffic; measuring field distributions; monitoring audio, chemical, or proximal conditions rather than just the stationary scene.

Thirty years of theory and applications in GIS have established workable ways to organize so much geodata. A GIS uses a few very specific forms of location modeling. In particular, this industry has developed robust relational databases keyed by static spatial features.

For example, a feature such as a land parcel can be used to key any amount of nongeometic attributes, such as historical and tax assessor data, and these can be held in a city's public records. From a modeling standpoint, the contribution of GIS has been topology: the representation and validation of enclosure and coverage. Of particular interest here is the tessellation (polygonal coverage) of a land surface by area features such as census tracts, vegetation zones, or market radii. Other topologies, such as points moving on lattices (e.g., transportation networks) have proven valuable as well.

These models make these information systems much more than automated map-making equipment, which persons unfamiliar with them too readily assume is all they are. Maps are made in GIS, of course, but as dynamic, thematic reports based on specific queries rather than as static documents. From these, further information may be generated any time. Spatial analysis, such as the terrain description overlays pioneered by the landscape architect Ian McHarg, may generate new data as well.

These processes demonstrate a distinction relevant to other forms of location modeling. Much as generating any number of maps from a GIS database differs from using the software as illustration tools to draw one map, other applications of location models, such as representations of who is present in a building, may be used repeatedly for a variety of purposes rather than automating one solution to one problem.

Open, reusable spatial data has become the main quest of the geodata industries. Location modeling involves stewardship of spatial information. Consistency, accuracy, and the pedigrees of data require continual upkeep.

If economies of reuse justify investments in location models, then the technologies of distribution appear as the fourth fundamental aspect of spatial information systems. In the past, these systems were mostly used and available in public agencies, and the geodata industry carried out government work. In America, continued investment at that scale has produced a National Spatial Data Infrastructure. Meanwhile access to such resources has been broadened dramatically by the Internet. Specialized middlewares developed in the late 1990s allowed information to be served dynamically. At the same time, GPS data become practical enough for current position to be used in remote queries to central databases. This completes a circle. Information is taken from places to remote centers of compilation, architecture, and analysis, from which it is then sent back into the field to let people know more about where they are. Navigation is only the beginning of this. There is a lot more to the technology of place than knowing where to take the next right turn.

Operations

Against the backdrop of so much spatial documentation, there emerges a more specific domain of place-based digital models which do not just map, but also help people select, construct, and operate physical sites. From the scale of market geography down to the scale of particular architectural elements, location models of one sort or another shape configurations of physical space.

For example, global organizations have models for planning the location and layout of franchise shops. Among GIS professionals, the caricature case has been the siting of new Starbucks franchises, which occur with enough spatial frequency to constitute sufficient data for mapping some demographic conditions. This case represents what has become large industry in spatial analyses. At one level, this is an instance of the geocoded target market industry we have already considered, but at another level, and due to its serial occurances, it is an instance of a location model of operations having architectural consequences.

As cultural geographers are fond of observing, when enough people and organizations use geocoded demographics to inform their site selection, they create a positive feedback loop in geographic distinctions, which functions as a giant sorting mechanism. For example, The Potential Rating Index for Zip Markets (PRIZM) data system, in use since 1974, has sorted Americans into forty "lifestyle clusters."[9] These demographic categories and their built consequences have become too deeply ingrained in geography itself to be abandoned overnight for alternative Internet schemes of business-to-customer relationship.

Market data models also account for the layout of individual establishments such as big-box retail stores. Terabytes of point-of-sale are collected and mined, and the results are used to tune the placement of wares. Sales per square foot, which is easily and frequently measured, has become one of the most powerful factors in retail planning, which in turn has become one of the most powerful factors in urban design.

Wherever facilities involve some kind of operations manual, still more detailed location models are likely to apply. Consider the science of flipping burgers, in which each franchise has a three-ring binder full of corporate intelligence, a document that is already a work of information design describing the use of so many pushbutton interface designs.

Location models also exist for longer timeframes, such as year-to-year facilities management. Digital building models used for design and construction are often adapted for models of leasing and maintenance. In more sophisticated instances, such models help represent the costs and benefits of operating a building over the full life cycle of its existence, and that helps justify increased investment in its design.

Not only the position but also the movement of people, goods, and services themselves becomes an application of pervasive computing. Much as an overnight courier can track a parcel, other networks can identify the location of mobile elements and the status of motion channels, to the degree where operations researchers may apply routing and scheduling algorithms to them.

Construction itself thus benefits from pervasive computing. Conventional building construction has involved a higher labor component with more waiting states, and lower-quality components with less adaptability, than most other endeavors with such a scale of costs. To mitigate these problems, a typical job site now uses extensive mobile communications, webcams,[10] programmable instruments, critical-path scheduling systems, etc.

As demonstrated in other industries, smarter assembly is not just a question of process management but also of better components, better software models, and more versatile design practices. For example, "mass customization" seeks to make unique components affordable by generating them from adaptable software models. "Design for assembly" seeks to reduce overall costs by emphasizing simpler processes rather than simpler parts. This is an example of how design is more useful when it is optimizing a system rather than adding features to isolated elements. "Product modeling" attempts to build standard hierarchies and libraries for managing and distributing design information. This has been essential in assembly industries that have huge parts inventories, such as aviation. Then beyond desktop endeavors at component-based design and management, the physical components themselves, as delivered to the construction site, can be digitally tagged with schedule instructions, installation specifications, and tracking information.

In general, and whether or not they are integrated with the design and construction of building components, operational models extend the role of built environments. Relative to the globalizing notions of network computing, then, this new emphasis on physical siting and ad hoc networking turns the focus back on the local. Smart operations involve more versatile access to regional infrastructures. That access can take many forms, from the scheduled use of utilities to automatic calls for maintenance to individual queries for available production services. What is important is that networks become perceived, not just as wires between computers, but as constellations of services, many of which connect an organization to its physical vicinity.

Environmental Systems

In a sense, any engineered system that monitors a physical measurement in order to regulate the state of its physical environment is already an instance of physical computing. A simple float valve in a tank is an analog computer of a sort, albeit one with few states and no memory. A more complex control system involving electronic communication between distributed sensors and actuators can probably be programmed by calibrating its component devices. Just about any assembly line requires such tuning. In the history of industrial engineering, analog numerical control predates digital computing. The idea of programming by demonstration can now be extended from physical machines to electronically regulated spaces. Control system devices have been natural candidates for the introduction of microprocessors.

Feedback control systems exist in the home and the office, but they have been especially critical for industrial process engineering. Automated production as we know it would be impossible without mechanical sensors and actuators. If factory floors are not thought of as smart places, that only illustrates our willingness to take so much technology for granted. Production facilities have been the focus of some classic works on implicit location models, such as Shoshana Zuboff's *In the Age of the Smart Machine* from the 1980s and Norbert Wiener's *Cybernetics* from the 1940s.[11] Sensors and "proprioceptors" were important in Wiener's biological metaphors for industrial process management. This anticipated environmental computing. Industrial realities commonly led to science fiction in which automation deftly intervened on humans' behalf. Modeling, implementing, and living with deterministic automation has been another matter, however. Current moves toward open systems and ad hoc aggregations in pervasive computing suggest a different mentality altogether.

For example, engineers' pursuit of interoperability occurs more at the level of systems than of the features of individual components. The mathematical models of system states cannot always obtain determinacy; incidental links and concurrent processes provide modeling challenges. The stability or responsiveness of a system may improve with the use of devices with lower individual performance specifications. Precisely optimizing an individual component may have little or even a negative effect on a system that uses it. Tuning and configuration become essential to the engineering process, and models must be created in a manner that allows for them.

Perhaps the most common example of a shift from determinacy to open systems by the application of more embedded intelligence has been the climate of buildings. There the quest for determinacy has given us hermetically sealed environments, recognizable by inoperable windows, which few of us enjoy. Where models have lacked detail or accuracy, and especially wherever human participation in tuning has been denied, there has been an enormous waste of resources. Mitigation of this stupidity has been the usual meaning of the phrase "intelligent buildings."[12]

According to many industrial ecologists, the life cycles of operated environments present one of the best opportunities for adding value by design. Since the 1980s, digital control systems have provided some improvements in zoned regulation of building climates. Now as energy waste becomes more topical again, the adaptability and resolution of these systems increases by an order of magnitude. When, over the life of almost any building, cumulative energy costs eventually rival the initial construction costs, investments in environmental computing are easily justified. Improved communication with external resource systems can schedule resource supplies and demands more efficiently, for example. This can improve overall utilization and life-cycle efficiency while at the same time allowing more flexibility of local control. For example, if you can focus light directly on an activity when you need it, overall ambient levels can be much lower, and the life-cycle savings will repeatedly repay the cost of the more intelligent system. Similarly with temperature, diversity can be more satisfying than sterile uniformity; many architects know this as "thermal delight."[13] Providing such tacit satisfaction proves to be a good case of technology staying in the background. Figure 5.3 illustrates some concepts of smart offices.

Beyond efficiencies in resource management, better models of intelligent buildings also seek to support the effectiveness of the people who use them. This brings architecture back to work practices, and it brings interaction design to a concern for physical context. Architect Frank Duffy has explained this distinction between "efficiency and effectivity" as the way to realize the benefits of information technology in the built workplace.[14] Because a payroll usually exceeds facilities as the greatest operating cost, designing for human effectiveness pays off more than designing for efficient facilities. Least common denominator spatial efficiencies of the sort made notorious by the comic strip character Dilbert save only on space. Marginally more expensive spaces that provide more casual adaptability of configurations, communication, and microclimates can yield greater gains in productivity and participation.

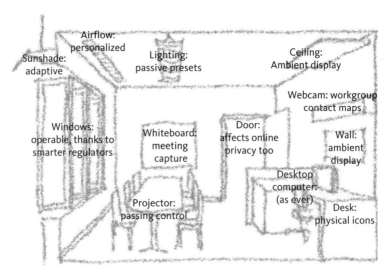

(a)

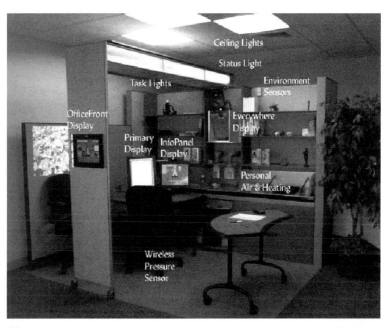

(b)

5.3 The smart office as more than a fast network connection: (a) Some general concepts; (b) BlueSpace, a prototype by IBM and Steelcase. (*Courtesy of IBM Research.*)

According to one of the more comprehensive early descriptions, buildings may be called "intelligent" mainly on the basis of how well their systems integration gives them adaptability. A response to changes in the weather is in some ways the least of these of these processes. Adaptability to organizational changes proves more valuable. Lack of adaptability is perhaps the chief reason why corporations underinvest in buildings.

At its heart, this smart buildings philosophy challenges the assumptions behind the reduction of built space to a bland commodity. The incremental cost of fitting out more adaptable, diversified space is offset by the greater benefits of presence, collaboration, identity, etc. that 1990s management experts identified as keys to the shift from industrial chains of command to more prosperous, and agile, networked businesses.

The challenge of creating useful location models reflects this larger trend. Beginning (but not ending) with the workplace, more effective (versus efficient) design depends on better resolved information about who is present, and what they are doing.

In application after application of feedback systems in a physical environment, the introduction of processing, memory, and diversified sensing creates a need to consider how people actually deal with, and not merely learn how to operate, so much technology. As environmental software supplants the crude logics implicit in mechanical control systems, its relationship to the protocols established by physical architecture becomes more essential.

Users, Actors, and Stages

In models of who is present, the best metaphor may still be that of the stage. Like architecture, location models for pervasive computing encourage role playing in specific contexts. The process of tagging and modeling users most truly distinguishes pervasive computing from both desktop computing and from previous environmental technologies. This work is fraught with technical and social problems, but it is essential. It is almost a tautology that better modeling of persons in contexts is the best way toward more human-centered design.

As an example of the ethical problems raised by user models in pervasive computing, do we condone the use of alert systems to detect whether a senior housing resident has opened his or her medicine cabinet on schedule? An elderly person's house can summon help if no motion has been detected for an unusually long period. In one "independent-living" facility that received architectural press coverage in 2002, the University of Virginia's Medical Automation Research Center helped outfit a building conversion project with noninvasive sensors used not only to detect particular events or their absence but also to create daily activity records from which to make more effective predictions.[15] Although such "profiling" worries many humanitarians, personalization in its many forms is nevertheless seen as an important near future in computing. So far, software personalization has emphasized modeling "the user" in a relationship of many tools to one user in one place—typically the desktop computer—rather than to many users or many places.

Now the spread of affordable tagging and sensing changes that outlook. Instead of personalizing one person's one window on the universe of the net, now the emphasis turns to interoperability of multiple systems in designated contexts. Often these are fixed to physical locations. Badges function on company premises. Radiofrequency identifiers work within a fixed radius. Official identities may have scope only within zones or organizations.

Location models provide a bridge from these tools to the scene, from the components to the scenario, from the activity to the identity. Who is here and what they are doing may be inferred from the tools they have brought along in combination with the tools they find on site. Conversely, declared identities can be used to personalize tools on site.

This is fundamentally different from the desktop. Quite often local models must infer or attribute actions. Who has control of a device right now, or who just said something? Much evidence for the centrality of perceiving persons comes from the problem of passing control. Like the speaking stone in a traditional lodge gathering, something must serve as a token.

Rooms, literal or metaphorical, figure in this process. In an Internet chat space, a room is simply a set of listeners. Rooms frame groups and offer them control of particular ambiences. Institutionalized sites, whether official or unofficial, suggest useful protocols.

Location models in software seem distinct from the architectural equipment of previous systems in their capacity for recognition and memory. Applications in personalization and authentication that now seem routine were not part of buildings less than a century ago. For purposes of matching protocols to whomever is present, software is likely to work better than hardware. On top of the fixed configuration of social relations formed in space, we get a more customizable layer of settings and communications.

All this tends to emphasize how users become actors (figure 5.4). Here is an example of how computing is about letting us take part in shared representations. Achieving the right balance in this process is the goal of interaction design. The usual objections include too much active operation, being stuck with conditions programmed by someone else, and having to learn settings that are not stable or permanent enough. The interaction designer's role implicates architecture in addressing these problems.

In order to maintain a representation analogous to actors on stages, many smart spaces use software agents. This makes it easier to maintain a continuous identity for participants in a space. As a semi-autonomous piece of software, an agent can move about more easily than the person or device it represents. It could perform errands on the Internet at large. This prospect received a lot of attention in the early 1990s. More pertinent to the issues at hand, an agent could circulate

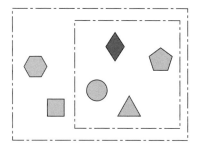

5.4 The fundamental technical problem: actors on stages

on a much more localized system, based on zones of a local area network, or preferably among a set of nodes whose member all shared some perimeter in physical space. Agents are intrinsically peripheral. More than any other component, they can stay out of the way under normal conditions and yet quickly come to the foreground when something needs attention. An agent can also represent a process or a set of project data that follows its author, so that the author does not have to go to one specific place to work on it. An agent is a representation of intent.

Open Systems for Extensible Places

For architecture's new digital layer to be effective, it must be able to accumulate in unpredictable ways. Pervasive computing must be extensible, and it must help make physical places adaptable. This extensibility must be casual; learning and configuring the technology cannot place too much burden on its users. Where that casual extensibility becomes a part of the psychological, social, and historical processes by which people come to identify with locations, the technology is more likely to be a success.

Casual access and random extensibility require much more elegant location models. A chaotic galaxy of independent systems replaces an existing monoculture of interactivity. Some of these systems aggregate and link up more deeply and enduringly than others. Some of those aggregations become systems in their own right— islands of intensified interactivity—each with their own local protocols and possibilities. It is these which require modeling.

Some technologies document and manage the places we already have, and some extend and transform the architectural types we already know. It is these which require modeling first. An appreciably short set of situational types becomes the basis for a diversifying field of interaction design. Location models should develop around these. Think about how they could do so with increasing commodity, firmness, and delight.[16]

A model represents a theory.[17] It is a perennial failing to mistake the representation for the reality, however. The map is not the territory. A representation shows this, not that, and does so with some particular purpose.[18] A representation may be about something that does not exist. Two representations can appear alike and yet refer to different things.

The philosopher Hilary Putnam has referred to the latter phenomenon as "the contribution of environment."[19] Because perceptions of environment involve discoveries, entities initially thought to be similar may turn out to be different. This condition is particularly acute when a describable process is used to model a complex, unpredictable system.

The ability to simulate has been one of the primary intellectual advantages of computing. Software allows more dynamics, complexity, variability, and visibility in symbolic representations than was possible before—enough so that some people overestimate its completeness. The analogy between a software process and the reality it is intended to represent is always at issue. Moreover, some complexities of the model may be no more describable than those of the corresponding reality.

As open systems become more prevalent in computing, and as the processes they model include more social complexities of everyday life, the challenge of designing them becomes much more like architecture. Architecture intentionally configures space by means of schematic proposals. It models form and identity more often than behavior or performance, and addresses all of these without the luxury of predictability. As a set of intentions about arranging space, architecture intrinsically applies a theory. As a philosophical means of ordering many aspects of experience, a particular way with space reflects a particular culture. This fixed configuration provides a framework for the flows of resources, authority, and people. Architecture thus represents.

Now the configuration of pervasive computing requires similar levels of theory, models, and intentionality. As an open system, this architecture of responsive places builds from many partial models of locality and situated action. Whether we appreciate its results may depend on how well we are able to represent environmental knowledge. This is a cultural challenge. Local aggregations of information technology represent cultural intentions.

6   Situated Types

Technology might at least distinguish among the life patterns that it so often alters. Designers of digital technology need to recognize living situations amid the legacy of conventions that they so readily declare obsolete. In a field better known for its frontier mentality, consider the role of typology.

Ubiquitous computing generally has been assumed to be portable computing, and the portable device has been assumed to be a solo device. But as communication devices repeatedly come into contact with one another, they create the need for protocols. Then it turns out that electronic data flows are hardly uniform or universal. Protocols become a whole-systems problem, in which the usability of an emergent collection of devices becomes more important than the features of any one. As we begin to internetwork mobile and embedded devices, we need more locational etiquette.

Technology still accumulates somewhere (figure 6.1). More richly mediated interactions generally require a set of supporting technologies that must be owned, housed, and maintained.[1] It is relevant that the great architect Louis Kahn once emphasized a distinction between spaces that serve and spaces that are served. Spaces that serve absorb their era's new technologies, and they are reconfigured but not replaced by these. The portable gear we bring along now serves too, and its interaction with the fixed must merit special interest, but the fixed arrangement remains the deeper half of the design.

6.1 Technology piles up

These settings accentuate the social aspects of information technology. Identification remains essential to place, to belonging, and to trust. This is highly subjective, of course. People tend to identify with settings they have casually appropriated, such as some corner of a park where they go to exercise—and not, by contrast, with settings that monitor, control, and foist a guaranteed experience on them. They are also more likely to identify with recurring experiences, especially those that result from social choices. Thus as technologies broaden choices in life, designers must address the source of the insiders' trust. Some scenarios of pervasive computing have fallen far short of this trust, as expressed in the common fear that devices are watching and talking to each other about us.

For these and any number of other reasons already explored, human interactions continue to exhibit categories, strata, and patterns. Such recurring configurations are natural; just about any species has them. Contexts remind people and their devices how to behave. That framing has often been done best and understood most easily as architecture. Something about the habitual nature of an environmental usage gives it life. Like device protocols and personal conduct, architecture has been a form of etiquette. Like most etiquette, architecture exists not out of pompousness, but because it lets life proceed more easily. Situated computing extends this age-old preference, whereas anytime-anyplace computing does not.

Typology

In short, we need a typology of situated interactions. By extending living patterns of inhabited space, we can strive to make technology simpler, more adaptive, and more social. The alternative is chaos. Much as free-form experimentation with unprecedented technologies in modern building often led to socially detrimental results, now pervasive computing creeps toward huge design failures.[2] Expect wrecks.

Recall that as a design philosophy, typology recognizes how creativity does better with themes and variations than with arbitrary innovation. It provides a framework for convention and invention to temper one another. Between conformity to a one-size-fits-all design and the chaos of infinite combinatorial possibility, there is a manageable range of recognizable situations. Design seldom benefits from infinite possibilities. It is more likely to be beneficial and appreciated when its variations occur on a few appropriate themes. Much as music finds richness in endless play within a relatively small number of specific genres, so now interaction design turns toward a less than infinite set of combinations. Much as architecture depends on habitual patterns for its livability, so the ad hoc local networks of devices must produce recurrent types. Whether those types reflect technological possibilities or human patterns becomes a matter of design.

As design participation broadens in digital technology, architects should awaken to these issues. As in architecture, where arbitrary freedoms often yielded dysfunctional spaces, type-defying free-form configurations will mostly not work in digital environmental technologies.

With the benefit of a longer historical perspective, architects understand how types are morphed, extended, and only occasionally made obsolete by new layers of technology. They have confronted dematerialization, and they know that all things digital will neither replace the built environment nor allow anything to happen anywhere in it. At a practical level, architects understand a component-based approach to designing macro-scale environments. As a scholarly discipline, architects have explored patterns of human activity independently not only of computers but also of buildings. The same cultural sensibility that finds substance in quiet architecture now turns its attention to situated gear.

Here follows a rudimentary typology of thirty situations (figure 6.2). These are by no means definitive and, as in most categorizations, could easily have been classified some other way. They could be organized by types of spaces, by types of technology used, by types of social conventions, or by types of activity.[3] All would interrelate. All would provide a reasonably short list. As it is, the groupings used reflect the usual categories of place: workplace, dwelling place, the oft-cited "third place" for conviviality, and an ever-increasing "fourth place," as it were, of commuting and travel. Given the emphasis on interaction in context, activity seems the right way to group these situations. A map of activities says the most about the usability of technology-modified places.

(at work...)

1. Deliberating (places for thinking)
2. Presenting (places for speaking to groups)
3. Collaborating (places for working within groups)
4. Dealing (places for negotiating)
5. Documenting (places for reference resources)
6. Officiating (places for institutions to serve their constituencies)
7. Crafting (places for skilled practice)
8. Associating (places where businesses form ecologies)
9. Learning (places for experiments and explanations)
10. Cultivating (places for stewardship)
11. Watching (places for monitoring)

(at home...)

12. Sheltering (places with comfortable climate)
13. Recharging (places for maintaining the body)
14. Idling (restful places for watching the world go by)
15. Confining (places to be held in)
16. Servicing (places with local support networks)
17. Metering (places where services flow incrementally)

(on the town...)

18. Eating, drinking, talking (places for socializing)
19. Gathering (places to meet)
20. Cruising (places for seeing and being seen)
21. Belonging (places for insiders)
22. Shopping (places for recreational retailing)
23. Sporting (places for embodied play)
24. Attending (places for cultural productions)
25. Commemorating (places for ritual)

(on the road...)

26. Gazing/touring (places to visit)
27. Hoteling (places to be at home away from home)
28. Adventuring (places for embodied challenge)
29. Driving (car as place)
30. Walking (places at human scale)

6.2 One set of situational types

Deliberating (places for thinking)

A thinker dislikes interruptions, and so knows when to close the door. Prior to the existence of electronic communications, this simple architectural tactic was sufficient. The door was necessary because this solitary space was preferably linked to a complementary, more public space for sharing results and deciding on courses of action. The classic metaphor is "caves and commons." Whether monastic cells and churches, or corporate cubicles and conference rooms, the relationship remains the same.

Communication technology obviously upsets this pattern. The telephone can ring behind the closed door. Receptionists become gatekeepers of incoming calls as well as the doorway. Outgoing calls can be useful to deliberation; one can acquire missing information without leaving one's desk. Larger volumes of calls can be pooled. Long before the use of cubicles, it became a standard image of early twentieth-century office work to see a group of workers sitting at a row of desks, making telephone calls all over the city. This was neither truly cave nor common, but an intermediate condition of cooperative work. (Meanwhile the solitude necessary for reflection could be linked to the commons remotely. If one had an idea while out walking around town, it was now possible to phone it right in.)

Today when different windows on one's screen constitute work and play, solos and collaborations, and privacy or its absence, the tactics of establishing solitary space have become much more complicated. Meanwhile the means of leaving the workplace to find that solitude have become much more practical. This is one reason why the "fourth place" of mobility has become more significant. To address the fixed workplace, where accumulations of technology are generally more sophisticated, "groupware" or collaborative systems have become a substantial domain of the software industry. Now pervasive computing adds a hardware and environmental component to those systems. With regard to address the basic need to close the door and think, Bill Buxton's simple "door mouse" was a significant early demonstration of situated computing. This real-space hack mediated privacy among members of a work group who were made peripherally aware of one another by their agreement to share video camera feeds among their offices. When the physical door was closed, this actuator sent a signal to the computer to stop the outgoing webcam signal.[4]

Presenting (places for speaking to groups)

Archetypally, the organizational common consists of a classical rhetorician in an oratory hall. Here the metaphor is to "have the floor." Some cultures provided a token of that status in the form of a speaking stone, the holder of which was to have the floor as long as desired, plus the choice of who would get to speak next.

Often the stage would include a proscenium to define the floor better. Similarly, the portico of a building represents the political authority from which the potentate speaks. Thus the Sultan of Turkey was known as the *Porte*. Alternatively, opposing benches such those as in a parliament indicate the two-sided nature of debate. A courtroom lets several parties observe. So does a theater in the round. The pulpit raises the preacher above the flock.

An ordinary conference room offers digital extensions that are remarkable by the standards of just a decade ago. Touch-screen controls embedded in a podium can switch between racked equipment and private laptops as projector inputs, control lighting according to stored configurations, patch in remote teleconference signals, and more.

Pervasive computing extends those possibilities and overcomes weaknesses of the bullet slide format by empowering the participants in a rapidly reconfigurable room. The holder of the stone becomes the controller of lighting, audio, and recording devices. A remote participant gains some leverage by which to get a word in. A meeting may create a new document (and not just the minutes) rather than recite the documents conceived separately by the participants. Although meeting ware has not yet found a widespread market, its research prototypes, such as Kumo Interactive at Fuji Xerox, have become quite robust.[5]

Collaborating (places for working within groups)

A worker's knowledge resides in a community. The middle ground between the thinker and the speaker has been increasingly instrumentalized in collaborative systems. In support of postindustrial management practices, these systems apply information technology to facilitate casual exchange rather than to proceduralize isolated tasks. Like architecture before it, network computing helps organizations learn.

These principles underlie the kind of technical work that appears in the ACM conference series Computer Systems for Cooperative Work (CSCW). Largely unknown to professional architects, these conference proceedings have developed a substantial literature on the design, ethnography, and technology of the workplace. In support of this growing field, the National Institute of Standards and Technology has developed a set of resource links[6] as well as a an integration platform standard for modular design research.

Dealing (places for negotiating)

As mentioned earlier in the discussion on the corner office, people prefer face-to-face conditions, especially for the exchange of power and opportunity. Recall that this sort of arrangement provides alternative configurations for the play of interpersonal distance. Again, precomputational technologies transformed the type by increasing its dependence on information support. Telephones summoned not only information but also additional players. Convenient printing technologies increased the use of graphics in support of decisions.

With the introduction of computer networks, some kinds of dealing places, such as stock trading floors, did dematerialize into information spaces, and those that remained were transformed by an increase in data displays.[7]

Today the conventional trading floor is a digitally enhanced operation that is typically rebuilt every few years. Here is a place where having the best information a little sooner is an important advantage. A trader is often equipped to mine intelligence from monitored markets around the world and then to communicate quickly, locally on the floor.

Documenting (places for reference resources)

Just about any work involves some information about what is here and what needs to done with it. Deals depend on documents. Equipment comes with manuals.

In recent decades, technology moved more of that information off the page to the sites of its use. Consider how many objects have instructions printed or glued onto them. Pervasive computing increases the value of such locally convenient documentation many times over. This is largely a matter of timeliness. A smart tag can furnish instruction in response to present conditions and linked sites. Objects with documentation be read with a portable device. Much as a building inspector has a probe to hold up to a wire to check for current, so many other kinds of inspection can involve some kind of local physical query.

One favorite model of embedded documentation uses wireless communication and position information to give a person documentation about an environment in which he or she is operating. One immediately practical application is repair manuals. For example, researchers from Carnegie Mellon University worked with DaimlerChrysler to demonstrate a position-dependent wearable computer by which transit repair workers could communicate with a remote expert at a help desk. The system was applied on the highly downtime-sensitive people movers at the local airport.[8]

As many designers like to speculate, located documentation could be implemented as an "augmented reality" overlay based on a comprehensive model. The often-cited Architectural Anatomy project at Columbia University demonstrated the early rudiments of this possibility by modeling the bare structural system of a building and mapping appropriate views of it onto transparent goggles worn into that building.[9]

Officiating (places for institutions to serve their constituencies)

The authority of an organization is exercised in the affairs of others. Such activity becomes the basis of institutions. Sites that have become institutionalized are among the most strongly typed in the built environment. Agencies, bureaus, and commissions maintain visible edifices and service counters. To physically visit one of them is to assent to its authority. Judicial systems carry this to the highest (and most enduringly built) formality, for instance.

Technology allows some officials to transmit their authority remotely. This requires authentication. The wax seal imprinted on a paper or parchment document accomplished that for centuries. Today we are asked to remember numerous passwords and personal identification numbers to serve the same need. As more and official business occurs online, only those institutional types where ceremony is valued in itself need a physical site. A church is the foremost example.

Official policies also influence access to network services. Such governance exists at the scale of buildings, cities, and regions, and for purposes of security, infrastructure planning, and the offering of online public services, for example.

Crafting (places for skilled practice)

Not all work consists of business letters, spreadsheets, and databases. Besides using documents, work centers around artifacts. Digital productions that reflect some of their history and the talent that went into them increasingly restore dignity to the economic arts known as crafts.

Context has been vital to craft. The settings of shop and studio reify work practices. Props, supporting tools, and work process configurations embody intellectual capital. They are prime instances of effective periphery.

Physical computing increases the potential of digital craft. Haptic interaction devices promise to restore touch to work in ways that the crude mouse only begins to hint at. Research in this work has expanded rapidly over the past five years. In one instance, a team at Interval Research and MIT coupled force-feedback to smart tagging for an input device that satisfies physical predispositions for continuous action. As the research team observed, "The natural world is often composed of and perceived as an infinity of continuums. People who have grown up within it are accustomed to moving and deforming and creating new possibilities from malleable media.... A button makes something happen automatically. A handle involves participatory control. It couples you to the environment. You are a part of a feedback loop."[10] By loading different force-feedback programs onto a handle, the device can be reprogrammed for different users and levels of expertise, for different applications, and to avoid using a mouse to fetch commands that have been buried in menus.

Associating (places where businesses form ecologies)

Places of work tend to cluster in particular where artifacts are traded, or where tacit knowledge resides in a community. The urban archetype of this is the district.

Transport and communication technologies eliminated the need for proximity but did not do away with clustering. Physical clusters could become larger and more specialized when people could drive to them. Now a district is defined as much by its architectural types as by any differentiation of wares: loft district, big-box district, etc. Our larger categories of domesticity, conviviality, and travel establish districts in themselves. Yet the single-use zoning so characteristic of the industrial city (where noxious factories and markets had to be kept away from residences) interferes with clustering. Districts must interweave and overlay.

A lot of this just happens on-screen. When booking a flight you can also book a hotel, look up a club, and send a greeting ahead to friend. More still happens on the back end. Online business-to-business reconfiguration (e.g., for bidding mechanisms, as at FreeMarkets.com) is largely about finding more natural associations. High-volume online trading sites tend to be far more credible as real network places than their counterparts in online socialization.

Learning (places for experiments and explanations)

While the global network, interactive publications, and online collaborations have introduced new channels of learning, about which much is made on alternatives to the in-residence university, those with the means to do so still prefer to sit and listen to the master (and then sit on the school steps after the lecture to take him or her apart).

In the review of activity theory, we have seen the importance of context to learning. Although the subjective aspects of engagement have become more important to interaction design, nevertheless the objective contexts may be adapted as well. Under pervasive computing, settings themselves learn.

This prospect has precedents in industrial technology. Programming by demonstration had a brief surge in the mid twentieth century as the need for numerically controlled machinery preceded the computational means to program it. Live demonstrations by machinists were recorded onto paper tapes by which the actions could be replicated. But that was all. Physical computing adds logic, memory, more precise and diversified sensing, and user response. Applications range from the elementary school classroom to highly specialized processes such as surgical training.

Learning flows in a two-way relationship between subject and context. Software that recognizes the abilities of respective users adapts itself to them, and guides their learning.

Cultivating (places for stewardship)

In addition to trading and making things, work also consists of stewardship. For example, environmental monitoring has grown into a multibillion dollar business. In one early instance of pervasive computing, in 1999 the city of London installed the world's first system of radio-linked, lamppost-mounted pollution monitors. These "streetboxes" used embedded computing to log various undesirable gasses against ambient conditions of light, temperature, and humidity, and to make intermittent radio connection to a nearby data collection center.[11]

With an increase in the number, connectivity, and programmability of practical sensors, environmental management can move beyond damage control toward a more proactive stewardship.

Watching (places for monitoring)

People fearfully assume that most initial applications of pervasive computing will be in civil surveillance. In reality most smart monitoring is not directed at citizens. For example, in an application known as geodetic monitoring, strain gauges and meterological sensors can be coupled to positional data and linked to software models to detect potential structural failures of bridges, mineshafts, or dams. Many other kinds of equipment and capital goods employ some form of monitoring. Computer vision first became widespread in applications to industrial robotics. Feedback control systems have been the primary application of embedded computation. The more processing, memory, and communication that the components of industrial process designs contain, the more these systems become stable, adaptable, and economical.

But then those remarkable powers of inference appear in human contexts as well. The social problem of production tasks being monitored, which is at least as old as the assembly line, now reappears in digital media at the level of keystroke counts and file content searches. It also becomes more remote and less visible, and it spreads beyond the workplace.

Surveillance has been accepted in countless applications, many of which have been cited in this text, yet remains objectionable in principle to many people for any given new instance. In a litigious culture that no longer accepts common sense as a form of prevention, we have become accustomed to surveillance in the most mundane of circumstances. Now as geopolitics presents more substantial risks, we face a disturbing prospect of losing any remaining social restraints to this monitoring. That prospect is enough to make many people resist the general notion of pervasive computing. Unfortunately that resistance may retard the development of benign applications more than the development of these detrimental ones.

At Home

Sheltering (places with comfortable climate)
Few archetypes run as deep as the hearth at the center or the roof overhead. Few aspects of domesticity have been so fully shaped by technology as the control of climate. Mechanical heating and cooling systems came along relatively recently in the history of architecture. Electronic feedback control systems have increased the sensitivity of environmental controls. The building thermostat would be a marvel of intelligent technology in any time but our own. Now an expansion in the number and capacity of sensor systems gives thermostats much greater sensitivity.

Recent developments shift the goal from conquest of climate to engagement with it. This is illustrated in the principle of "thermal delight." Some daily and seasonal change in indoor environmental conditions, and even in the use of space as a consequence of those changes, seems preferable to uniformity. Participation in such cycles remains one basis of being in place. Although some people will seek maximum convenience in uniformity, even at the cost of being cut off from climate, others will find satisfaction in programming their dwellings for more cyclical living. That is only natural.

Recharging (places for maintaining the body)

We go home to recharge. Other than sleeping, our acts of recharging depend on domestic infrastructure—plumbing in particular, but also kitchens, entertainment centers, hobby workbenches, and so on. Recharging involves solitary interactions with technology. It utilizes the "serving" spaces of the house—those more likely to incorporate smarter machines. Hence a European visiting Los Angeles might marvel at the obsession with a private bath for each bedroom in even the humblest dwellings, or at how the real estate agents recite the number of appliances in a house rather than any architectural features.

What happens when some domestic recharging facilities become wearable? This will be especially pertinent as the wearable or portable device relates to the fixed and serving spaces of the home. Wearable computer pioneers have recognized applications in body monitoring. Statistics formerly only available for hospital patients or to sports professionals now can be uploaded from privately wearable devices. Some of these devices might be worn outside the house, such as when jogging, but their accumulated data would most likely be read from the home.[12] And some would serve a house-bound patient. Perhaps the largest growth sector for "smart home" applications is in systems to permit the elderly to "age in place" in their own homes.

Idling (restful places for watching the world go by)

Passive entertainments dominate domestic life. At rest we watch others. Whether surfing channels, websites, or (traditionally) the countenances of people walking by, the flow of glances soothes. The restfulness of people watching underlies such traditional typological components as the porch, the balcony, the picture window. Traditionally these elements would open onto the street, where casual engagement with passers-by would make a set of such spaces into a communal living room. Entertainment technologies as simple as the Sunday paper disrupt this. People retreat into their separate worlds. It is hard to justify front porches today as cars rush by and people sit inside playing video games.

When each person entertains himself or herself in isolation, however, the "served" communal space of the house falls into disuse. Hence the recent trend for a "great room," which keeps family members in the same place, however separate their attentions, and which restores an arrangement long ago dictated by the common need to sit near a source of warmth. Clearly the type has morphed under technological change.

Electronic communications give the domestic idler a window on the world at large. The public spectacle becomes accessible from the convenience and privacy of one's own bedroom.

Meanwhile, escalating voyeurism has been evident in the recent mania for "reality television" based on constant camera feeds; in the fascination for recreational statistics, that is, in watching patterns in what people are doing and watching; and in a corresponding exhibitionism on the part of growing ranks of the population. That which is on camera is more somehow more real, even amid the privacy of the home.

Is this the front porch morphed beyond recognition? The technology transforming the type is the camera, in all its increasingly real-time and digital manifestations.

Confining (places to be held in)

Historically, domestic types have included various forms of confinement, whether voluntary or involuntary, mandatory or by default. Prisons, hospitals, asylums, and other such residential institutions must be included among these types. Some of these institutional stays can now be avoided with the use of a smart wristband, which allows an otherwise normal routine but calls a doctor or parole officer as needed. The introduction of information technologies transforms the prevailing conditions of each of these specialized residential types.[13]

Despite the astonishing number of people that it incarcerates in its prisons, America's largest ranks of the confined are the millions of suburbanites too young or too old to drive. High use of the Internet by the elderly has become a fairly well-documented phenomenon by now, but how does that use transform the use of physical space that many experience as a barrier? For example, would a senior gardener use a residence facility's community garden more often if he or she could watch remotely for the arrival of friends there?[14]

Servicing (places with local support networks)

The domestic web of serving spaces and incremental resource flows is not confined to the household. While in grander residences this domestic economy could be played out in the arrangement of serving spaces (e.g., the served upstairs, the servants downstairs), more usually the support network is provided by the neighborhood. The domestic scene belongs to a local economy of providers and services. The relation of each and all houses to the conveniences of the neighborhood is their basis of place.

The history of domestic technology suggests a continual democratization of service. For example, anyone with a refrigerator (and not only those with servants to do the daily shopping) can have fresh foods. At the same time, technologies of transportation have diffused the local concentration of services. By far the most important role of the automobile is the freedom it provides for random access. The pleasing reality of a tightly knit set of domestic services in a neighborhood is lost on autoabiding suburbanites, whose settings force them back into the car between even the smallest transactions.

Teleservice providers such as Internet-based grocery delivery services have tried to mitigate such problems, but so far without the economic success that was first expected of them. New social patterns both of working at home and spending more time traveling demand more flexible domestic economies. Unfortunately the market has not yet found many solutions.

A lot of goods and materials enter and leave the home little by little. Think how many groceries you carry into the kitchen each year, for example. Domestic economies are made up of these microtransactions, and regional economies are the sum of millions of these. Many forms of built space reflect some daily routine of dropoff and delivery. The better urban layouts often provide an alleyway or mews for such services.

Smarter technologies increase the resource efficiency of these many distributions. Several recent home automation projects in corporate and university research centers have addressed the domestic economy as something to be managed with an integrated information system (figure 6.3).[15] For example, if domestic control systems can schedule appliance jobs such as dishwasher cycles at off-peak hours of electricity demand, local capacity is utilized more efficiently.

On the Town

Many designers recite urban sociologist Ray Oldenburg's principle of the "third place": a location for conviviality that supplements home and work as a site of everyday life.[16] Designers who fear that technologies more often drive us apart than bring us together turn to social life for studies of assimilation. The ubiquitous, and often obnoxious spread of cellular phones into restaurants has been a harbinger of this problem. Thus we are not consoled by prototypes of smart glasses that tell waiters which tables need refills.[17] Much of the appeal of gathering over food and drink is that it is very low tech. Few of us enjoy having any more stuff in the way. Such social contexts may give pervasive computing its toughest tests of appropriateness.

Eating, Drinking, Talking (places for socializing)

Conviviality over food remains the best basis of the good life and of good places. The local bistro or pub provides an environment that is not available in the commercialized, customer turnover-inducing world of the chain restaurant. The goal instead is to have the best service with the least interference. Studying the protocols of waiting on

Activity recognition   Aware home   Georgia Tech
www.cc.gatech.edu/fce/ahri

Building performance   Intelligent workplace   Carnegie Mellon
www.arc.cmu.edu/cbpd/html/iw/iw.html

Linking from home   House_n   MIT
architecture.mit.edu/house_n/

Meeting support   Kumo interactive   Fuji/Xerox
www.fxpal.com/smartspaces/

Narrative playspace   Kids Room   MIT
vismod.www.media.mit.edu/vismod/demos/kidsroom/

Office systems   BlueSpace   IBM/Steelcase
www.research.ibm.com/MobileComputing/BlueSpace.html

Passive learning   Adaptive house   University of Colorado
www.cs.colorado.edu/~mozer/nnh/

Person recognition   Easy living   Microsoft research
www.research.microsoft.com/easyliving/

Trade magazine/convention/catalogue   Electronic house
www.electronichouse.com/

Training toy   Lego dacta   Lego
www.lego.com/dacta/products/

6.3 Ten projects of smart space

tables makes interesting ethnography.[18] Here it is evident that etiquette is a means and not a burden. Here too, the preferred role of high technology is peripheral, to establish ambience.

New media make it practical to provide a more diverse range of ambiences for social gathering spots. They also make it easier for patrons to identify and locate those locations. The question is whether this can be accomplished without the undue thematization that interferes with the social spontaneity that is the prime basis of the type.

In some exceptions, the technology may come into the foreground, and the theme may be welcomed as a part of the social game. Consider the recent prospect of the "image cafe," for example. Cybercafes have been socially centripetal; participants were drawn away from the physical scene through their respective screens. An image cafe reverses this in what is essentially visual karaoke. Images brought by patrons can appear in a sequence or a mosaic on shared table tops or on a common wall. Customers can unload their digital cameras for others to see their photos, or they can suggest uploads to a video disk jockey, etc.

Gathering (places to meet)

The act of rendezvous characterizes many urban types. It often occurs at a landmark, such as Marshall Field's clock on State Street in Chicago. Something easy to identify, remember, and find thus takes on an active social role. Visitors find landmarks essential in this regard. Regular denizens can meet just about anywhere, however. The habitually appropriated spot takes on social significance for them. For example, a particular group of teens may meet at the abutment of a particular bridge each weekend. Some older men might take over a corner of the park for lawn bowling.

Communication technology assists in physical gatherings. The first thing said on Alexander Graham Bell's phone was "come here, Mr. Watson." Less directly, advertising on broadcast media increases the importance of "destinations." When coupled with the reality that people tend to decide where they are going before they get in their cars, and tend to spend their money near wherever they get out of their cars, electronic inducements to meet someplace assume great social and architectural importance.

Here the mobile phone is a positive force. Decisions about gathering often occur on the move, and quite often in response to current conditions. Frequent back and forth calls amount to polling: "How about here?"[19]

Cruising (places for seeing and being seen)

It is a short step from polling to cruising: that is, social circulation in search of one's kind. In civil society the promenade formed a gracious opportunity for its participants (not necessarily so graciously, in their minds) to check each other out. Various urban types from the boulevard to the boardwalk to the grand stair at the theater have served this end. The history of technology has yielded new variations. For example, the railroads introduced the resort hotel, where the social cruising could be carried out over a week-long visit, with the assurance that only those of similar means were present.

Global travel has radically increased the diversity and frequency of passing encounters, and in the process has eliminated many formalities while at the same time sharpening the eye for social cues. Alas, many people are reduced to the signs on their shirts. Designer logos, sometimes literal messages, and predispositions about various "lifestyle choices" dominate the passing social encounter with a stranger.

Wearable computing devices diversify this play of signs and add processing and memory to its dynamic. A wearable lifestyle device can poll the vicinity for others sharing a selection of attributes—maybe just someone on the same sidewalk who is listening to the same song.[20] Marshall McLuhan predicted that electronic media would retribalize society. Now this is happening in physical space.

Belonging (places for insiders)

Tokens of belonging, that essential construct of place, thus spread out onto bodies and portable gear. What one wears has always served as a badge and a means of entry, of course. What changes is that formalities shift from appearances to information-technology handshakes. The spatial archetype is the gate. The forms include the perimeter, the gangway, and the velvet-roped queue. Gatekeepers now have many more ways to broker belonging. The criteria of insideness and outsideness are more numerous than ever before, and each person present maintains a partial status in both categories.

Smart identification cards or badges have more social appeal when they are used for two-way communication in social navigation. This could be more fun, or at least more variable, than signaling status by colors and logos on clothing.

Shopping (places for recreational retailing)

A lot of social belonging is expressed through shopping. Because it gets people out walking when otherwise nothing else does, shopping has become the glue by which public places are put together.[21] It is the driver and consumer of pedestrian activity. There is even a shop at Walden Pond.

Retail infrastructures combine the marvel of unprecedented reserves with the wonder of instantaneous flow on demand.[22] The products moving off shelves (sales per square foot) become the vital statistic of real estate value. Although traditional marketplace types endure, and are revived, other shopping formats are among the prominent typological inventions of the past hundred years. The influence of technology on these morphologies goes beyond escalators, strip centers, and the car as a shopping cart. For example, when specialty superstores replete with tens of thousands of items in, say, housewares, become spectacles and recreations, they can seem like temples to some god of digital inventory control. Their awesome inventories would be impossible without smart devices and embedded information. Some of these are as simple as the barcode scanner, some are as complex as personalized shopper databases.

Both an information layer and a tradition of personalized service surround the act of shopping. How pervasive computing helps customize individuals' physical shopping experience is of considerable interest to interaction designers today.

Once only the wealthy or the regular patron could expect a personalized response from retailers to include knowledge of their preferences and past purchases. Now the early adopters of digital technology have the lead in personalization. The ideas behind this have been implemented successfully on the Internet. Firefly, the prototypical inference engine for monitoring personal tastes, appeared in the mid 1990s, and just a few scant years later Amazon.com had become good enough at recommending books to a customer that it tended to suggest items that he or she already owned.

Pervasive computing brings personalization back into the physical marketplace. That milieu remains important, as demonstrated by the fact that the majority of purchases in malls are not premeditated—the longer people are physically present, the more they buy. Now the personal database is built and referenced in conjunction with the act of physically handling products. In some cases, this custom service adds the ambience so important to discretionary shopping. This is even more the case when the dialogue is furthered by the personal digital gear one brings along. The portable meets the embedded at the clicks-and-mortar venue of sale.

Sporting (places for embodied play)

Quite often the "third place" takes the form of a playing field. Sometimes it is out in the wilds. Locations for training and prowess belong to any typology of conviviality.

"Sportsmen" and other outdoor enthusiasts display a great proclivity for gear. No other people has been so highly equipped as Americans at their sports. Performance clothing, body monitors, instruments of tracking and timing, even a mere hockey puck can be high tech. For an everyday example of ubiquitous digital gear, the $30 bicycle computer shapes countless workout routines. Using an optical signal to count turns of the wheel, it calculates distances, velocities, and averages and derivatives thereof.

Sometimes the sites of sport can be equipped themselves. Imagine an urban sport made of tagging so many fixed spots downtown. People could play capture the flag using PDAs on the streets of TriBeCa.

Attending (places for cultural productions)

As more people incorporate aspects of art and entertainment into digital crafts, and conversely, as institutionalized cultural genres become increasingly influenced by digital media, the experience of attending a performance yields new types, particularly formats based on audience response.

In one of the largest public demonstrations of interactivity, MIT musician Tod Machover's landmark audience-driven Brain Opera was "performed" at Lincoln Center in New York in the summer of 1996. Its cast of thousands included musicians, a crowd, and remote online participants, and its three formats included an outdoor crowd-driven display, lobby exhibits of hands-on hyperinstruments, and a sit-down performance that was in part steered by remote audience on the Internet.[23]

In a smaller, more recent installation, the SenseBus project of the Art and Robotics Group in Toronto networked a set of simple microprocessors, of the sort found in household washers and dryers, into a space whose ambience could be modified by haptic interactions with physical objects. Waist-high metal cylinders sensed proximity and orientation among themselves, gave vibrating feedback about those as they were pushed about by visitors, and modified sound and lighting in their space according to their collective configuration. The next visitors experienced the state left by previous ones, and thus the ambience slowly evolved with each successive visit.[24]

Commemorating (places for ritual)

Despite so many innovations, many forms of conviviality seek ritual types in the built environment. The endurance of these types is the very basis of their appeal. Churches and other sacred spaces remain foremost among these. Other types for other activities considered devotional (sports increasingly among them) sustain environmental memory in the same way. To abandon these for anytime-anyplace arbitrariness might please no one. We frequent chapels, arenas, ballrooms, historic landmarks, and personal ritual sites because we want to. The more that any of these has been used in a culture, the more it seems "natural" to continue visiting it. The basis of typology is that the recurrent patterns of the environment serve appealingly as a cultural memory device.

The ways of the high-tech nomad have been expanding. What environmental types do they support?

Gazing/Touring (places to visit)

Tourists want to see, and be photographed at, known monuments. As described earlier, an induced demand for tourism feeds on clear landmarks, around which often only incidentally related resort amenities are built.

Given the millions of photographs that are taken over and over of exactly the same views, the camera has to be the key technology for these sites. The "overexposure" of particular sites suggests a steep increase in demand; people want to see what everyone else is seeing. Increasingly, tourists bring along information, often in the form of guidebooks, which shapes what they see and how they see it. Guidebook selections distinguish tourism market segments.

Some guidebooks can be loaded onto a PDA. Lonely Planet offers datasets by city that are about the same price as a printed guidebook. Updates can be downloaded in seconds. Contexts can be searched and sorted. Additional guides add no additional bulk or weight to one's luggage.

When used on a hand-held computer, guidebooks couple well to navigational systems. This is a matter of integrating content, not unlike the Access guides by which Richard Saul Wurman had made an early case for information design. Before the dot-com crash, Vindigo.com was delivering free local guides, reviews, and navigation to the Palm operating system for Boston, New York, Washington, Chicago, and San Francisco.[25]

Filtering positional data by interest turns a PDA into a specialized guide on, say, restaurants or architecture. In an often-cited early demonstration, the Lancaster Guide project implemented a context-sensitive tourist guide using radio communication to hand-held computers. This two-way link allowed users to make bookings and to query one another.[26] The precedent was set; how one filters available geocoded information becomes an expression of one's niche as a tourist.

Hoteling (places to be at home away from home)
For the global traveler to be always and yet never at home requires a complex web of support services. The primary concentrator of these is the hotel. Well-understood types of hotels differ not only in price and elegance but also in the sets of services they provide. A lodgings guide is a simple lesson in typology. There are fleabag walkups; convivial pensiones; atmospheric, small, family-run courtyard places; faded downmarket commodity boxes; antiseptic camera-monitored chain inns for the risk-averse; business-oriented suites with data jacks and ergonomic desks; full-service conference venues; sybaritic resorts; aristocratic retreats; and so on.

Adventuring (places for embodied challenge)
The traveler, detached from direct experience and deep knowledge of homeland, engages the Earth in other ways. These tend to be concentrated, innovative, recreational; in the favorite word, "extreme." Most forms of adventure sport and ecotourism depend on some sort of high-tech gear. So far this technology has mostly helped people keep out the elements; climb, float, or wheel on surfaces; and made camp cooking palatable. When wearables deliver information and communication, however, new forms of extremity become viable.

Driving (car as place)
Among technologies that influence settlement patterns, perhaps none has had greater impact than the automobile. Moreover, the car itself has become a place where many people spend much of their day. A typical car depends on several internal microprocessors to operate, and now research has advanced communications to and from the vehicle as well. Citizen's band radios and cellular phones already provide plenty of chat, for which they are rightly resented as a hazardous distraction to driving. Next come navigation systems; by mid-2000, Hertz had deployed its NeverLost system in over 30,000 rental cars. General Motors expected to have a million subscribers to its OnStar navigation service by the end of the year 2000. The auto industry has taken a liking to the rearview mirror as a sky-visible site for telecommunications. One maker, Gentex corporation, already has mirrors

with "GPS system interfaces, cell phones, microphones, emergency notification systems and the like" on more than a dozen vehicle models in North America."[27]

It would not be difficult to have smarter transportation. Losses based just on the measurable costs of everyone sitting in traffic quickly run to the billions. The Federal Highway Administration has therefore established standards for intelligent transportation systems, toward which companies such as iteris.com have begun to develop systems. The goal is to monitor flow, recommend trip times, and reroute traffic for better throughput on the road network. Major trucking companies tend to be the first to gain the benefits of such efficiencies, but individual motorists obtain some side effects, such as better roadway assistance programs. In some cases the road system itself is outfitted with software. For example, in 2002 the District of Columbia undertook a $100 million overhaul of its traffic lights, in which sensors in the roadbed and cameras overhead provide enough data to govern light timings more effectively. The deployed system is expected to result in less "blocking the box" and as much as 20 percent less waiting time.[28]

Walking (places at human scale)

For most of us other than the captives of sprawl, walking remains the most fundamental form of mobility. Walking gives scale. Interface designers in search of natural technologies might focus here. Technologies that we can use while walking are the most truly portable.[29]

Many devices are light enough to carry comfortably but are heavier in terms of usability; they require us to step out of pedestrian traffic, or even to sit down to use them. Many skills and perceptions change when one is walking. Some are more acute than when we are still and others are blunted. This simple distinction begins a typology of portable device design.

Consider how we walk in relation to others and in relation to fixed environments. Urban designers understand difference in pedestrian activity, and their work emphasizes typologies of form. Thus we come full circle to the persistence of repeating patterns as a manifestation of affordances and appreciations of embodied environments.

The difference between ubiquitous and situated computing appears vital. Ubiquitous computing as promoted by the information technology industries has mostly been a matter of pure mobility, with little regard for locally embedded systems. It has emphasized access to the same information everywhere. It has been geared toward connectivity 100 percent of the time for a few people, rather than providing information when useful for 100 percent of the people in a specific location. It has sought a one-size-fits-all solution for technological interoperability.

By contrast, situated computing is based on the belief that such universality is neither attainable nor desirable. This approach questions total mobility, advocates local protocols, recognizes forms of tacit knowledge, and taps into more kinds of embodied predispositions (figure 6.4).

At least as much as we need to connect to the same net everywhere, we need different places in our lives to help differentiate, structure, and facilitate our activities. How do connections between mobile and embedded technology adapt to our intuition of being in a place,

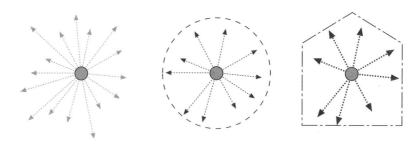

Shallow connections anywhere   Filter by proximity   Enrich by local protocol

6.4 Toward niche protocols in communication

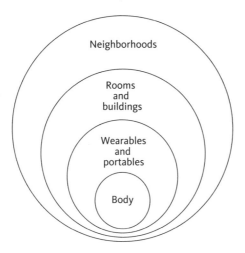

6.5 Scales of place

give us a sense of scale (figure 6.5), and how do they help us assimilate the conditions we find? To understand our places in an ever more mobile world, we must develop better patterns, protocols, ownerships, and trust in location-aware, situated gear.

Situational types suggest one provisional basis for designing this emerging digital layer of space. Since no universal standard can support local particulars of interactivity, differentiated categories of local technology must arise. Since not every situation can be subject to custom engineering and tuning, classes of adaptable technology must arise. Typological design should advance the appropriateness of location-aware technology more quickly than universal standards or one-time circumstantial configurations.

If the means are not yet clear, we may at least agree that the ends of design must include some approach to appropriateness other than solely technological features and their performance specifications. An open standard could establish practices for any particular type. This seems a matter of *protocols*.

Could there be a way to embed situational instructions into particular physical locations so that arriving mobile devices could properly configure their behavior while visiting? By crude analogy, physical space is full of text signs telling us what we may and may not do in their presence. This seems a matter of *tags*.

Many researchers observe that representing the human activities and merely recognizing the human actors that drive physical computing environments is perhaps the single most difficult computational challenge. How closely does this connect with the physical geometry (scale, configuration, boundaries, etc.—all architectural factors) of a situation. This seems a matter of *models*.

Within some typology, will particular technological arrangements recur often enough, with clear enough themes and variations, to become packaged into "sets"? (Recall that a television receiver was once referred to as a "set," for instance.) What would a driving set, a hotel set, or a set for city government public service counter be like? By the very nature of settlements, buildings, and cultures, we may expect some sort of typological distribution in recurring combinations of digital environmental enhancement. But the next step right now is just to build the first workable sets. This seems a matter of standard *interconnections*.

The usability of well-made traditional places now appears as a rich basis for design of context-aware technology. Whether it is organizational, social, or domestic, space awaits rediscovery for its richness of social framing. What architects refer to as vernacular or common typology should be of increasing interest to anyone studying located technology. Interaction design's increasing emphasis on context suggests a limit to the marginalization of building design and a new-found interest in periphery. This new work's emphasis on usability and appropriateness eventually leads beyond its immediate concern for the situated task toward a more inclusive configuration of the physical context. As a discipline, interaction design will need to understand how built space contributes to the conduct of life.

III Practices

7 Designing Interactions

The more that artifice permeates life, the more design becomes an essential liberal art. Because technology affects so much of what we do, even who we think we are, its design involves judgment and appreciation. Thus interaction design increasingly takes the form of a practice.

As a practice, design means more than making things look pretty, although good form is usually welcome. It also means more than making things usable, since something quite usable might nevertheless be useless. It does not flood the world with all technically possible gadgets and distractions. What we choose to build matters just as much as how it looks, or well we can make it operate. This choice is largely a social process; proposing what to do involves negotiation. Part advocacy, part virtuoso authorship, part ethnography, part engineering science, and part architecture to live by, interaction design needs conscientious multidisciplinary discourses.

Almost by definition, practices shun formulas. A practitioner does not need an explicit definition of design, and a theorist may never arrive at one. People who haggle endlessly over the meaning of design actually may not be seeking a unified science at all. Instead, the idea of design benefits from constant negotiation.

Strategic Legitimacy

The design of interactive contexts occurs within a larger cultural change, in which creativity has assumed much more economic value. When businesses themselves become objects of design, then design is good business. When successful business planning involves inventive schemas, it is fair to refer to it as strategy.

The word *strategy*, which has military origins, emphasizes cleverness in setting an identifiable course of action. It suggests that discovery may play as great a role as analysis in supporting such decisions. This kind of work demands interdisciplinary insights based on engagement of projects rather than specialization in subjects. It revolves around propositions rather than predictable solutions. Its fundamental act is to see conditions in some fresh way: "What if we use *this* like *that*." According to Herbert Simon's well-known explanation, design is "planned artifice":

Everyone designs who devises courses of action aimed at changing existing situations into preferred ones. The intellectual activity that produces material artifacts is no different fundamentally from the one that prescribes remedies for a sick patient or the one that devises a new sales plan for a company or a social welfare policy for a state. Design, so construed, is the core of all professional training; is the principal mark that distinguishes the professions from the sciences.[1]

With or without established professions as a base, a great range of disciplines work creatively and in effect professionally to plan interactive artifice. Design practices pay off most when they are incorporated across conception, development, use, and habituation of a product, service, or environment—and are not just used to correct the usability and identity of a product conceived by other means. Usability engineers, database designers, visual artists, business managers, architects, psychologists, ethnographers, and device developers all shape interaction design. In doing so, they draw on bodies of knowledge that vary in how codified they are and how understandable they are to others. They may never agree on *what* is design, but they know *when* is design.[2]

Almost any participant in the process raises objections (figure 7.1). Engineers and software developers, who tend to think of themselves as scientists, have trouble accepting this indeterminacy. Although they may appreciate notions of elegance as well as anyone, they have been conditioned to distrust how design is something outside the supposedly value-free experience of science. "What are the rules?" they ask, "How do you measure the results, and how do you know when you are done?"

By contrast, visual artists may cling to the image of the signature designer. To them, design remains an autographic process in which conception and execution intermingle. Methodology interferes with such aims. The pursuit of expression leads to unique works rather than products or systems. It also discourages the use of design alternatives, since for any project there can only be one proposal to which one is passionately committed. While there is a place for such art in all

Yes, But. . .

Design is a luxury	More appropriate systems are a necessity
Usability first	For beginners, masters, or communities?
Specify performance	Optimizing a part may disoptimize a system
Measure keystroke-level behavior	What about intent and involvement?
Procedures and automation	Participation and flow
Understand the information context	Also the physical and cultural context
Design has no definition	Constantly negotiating it is useful
Aesthetics are subjective	Measurement biases toward the measurable
Enough products and information	Richer contexts, please
We can imbed intelligence into everyday objects	Unless they are useful, and polite, they will still seem stupid

7.1 Design advocacy

things, this image of the designer does not earn much respect from disciplines that seek negotiable, reproducible results. It also discourages individuals who do not believe they have enough talent (or ego) to operate in this manner.

Business analysts meanwhile still question design's contribution to the bottom line. If until rather recently most businesses held little regard for design, that is because they saw it as something applied after the fact. When it merely dealt with packaging (including "front end" interfaces), design seemed superficial. When it was thought of as applied decoration, which may still be the most widespread connotation of the word, design implied cost rather than income. Industrial design's origins in corporate identity, in which a brand is applied to something that has already been produced, only reinforced this perception.[3] This is a vicious circle. When design is applied to productions that have long since been analytically conceived, then the self-fulfillingly unimpressive results can be used to demonstrate the superficiality of design.

Now that circle is breaking. Widespread computation makes business strategies based on reductive numerical models more or less equally available to everyone. Because efficiency models become more of a prerequisite but less of a competitive advantage, strategic emphasis shifts to design. The design of industrial products such as shoes and automobiles has advanced considerably as a result.

The design of businesses themselves has done so as well. As became a mainstay of management consulting in the 1990s, digital networks quickly allow organizational change to surpass task productivity as a benefit of information technology. By now it is common knowledge how recombinant communities of knowledge workers remake older chains of command. This postindustrial conception of design has promoted a more interdisciplinary approach to the building of information technologies.

In particular, an emphasis on communities of knowledge has legitimized more emphasis on context, even prior to the spread of pervasive computing. This is because knowledge workers do not follow procedures so much as expertly play their contexts.[4] Without an ability to improvise in context, people who are merely following official prescriptions are utterly lost as soon as they stray from known conditions, which of course happens all the time.[5] Redesigning organizations effectively depends on representing this kind of work.

When organizational change creates success factors that are cultural, then it requires more attention to design. When insights emerge more from the engagement of project particulars than from the methodologies of specialized disciplines, then propositions become as important as analysis. When "wicked problems" introduce controversy, complexity, tradeoffs, and lack of closure then deliberations replace specifications. When value deliberations occur over propositions in the face of indeterminate problems, then design is happening.[6]

A strategem is not an indicated solution, but a proposition. Design involves discovery; identifying and solving a problem may occur simultaneously. In some cases the propositions become better known than the problem they have solved or the process by which they are reached. Neither reductive analysis, nor blue-sky invention, nor necessarily corrective, the essence of design is practical synthesis. As IDEO founder David Kelley once put it, "The designer has a dream that goes beyond what exists, rather than fixing what exists."[7]

In sum, business strategists who use creative propositions to move beyond the limits of reductive analysis are legitimizing a widespread approach to design. "Designing will become a larger part, even the whole, of what a leader does,"[8] said John Kao, who has promoted the study of creativity at Harvard Business School. "This is the age of creativity because that's where information technology wants us to go next," is first among Kao's explanations of why this is happening now. "This is the age of creativity because of the primacy of design."[9]

The Need for a Craft

Valuable proposals are no more discovered out of the blue than deduced from reductive analysis; instead they arise from the committed practice to the flow of work. In contrast to the antitechnological folk meaning of the word, postindustrial form-givers refer to this flow as craft.[10] Like its preindustrial predecessors, abstract craft is inseparable from its contexts. Clients, projects, workplaces, and technologies all provide contexts for design work.[11] By addressing systems, services, products, and communications, designers seek to rearrange some context for the better. Experience with a particular set of contexts remains essential in design expertise. Work is not just something you do, but also someplace you go to each morning.

A practice cultivates mastery and judgment. Based on lifelong learning and devotion to a core set of knowledge and values, it has intrinsic benefit for those who take part in it. In that regard, a practice is a goal in itself. This quality is demonstrated by anyone who works primarily for the right to continue to practice.

The tacit knowledge and values built from such experience are expressed and maintained through "reflection in action."[12] This reflectivity is welcome; this kind of work has human benefits to those taking part. To the extent that it resolves conflicts over differing concepts of value, it has organizational or cultural benefits as well.

Practices evolve of course. Technology-centered interface design becomes human-centered interaction design. Emphasis on the solo task gives way to emphasis on social processes. Optimization for performance specifications or first time usability metrics gives way to whole-systems engineering for configurations we can live with, master, and tune. Goals such as these align design with both business strategy and cultural expression. They also provide a setting for digital craft.

Without such shifts in attitude, people continue to treat computing as more of a technical than a cultural problem. In particular, current interfaces illustrate how many computer scientists are biased toward efficiency with technological resources rather than human attention; or to put it bluntly, toward convenience for computers before convenience for people.[13] This orientation shows in a typical admonishment in Microsoft Windows for example: "The computer was not shut down properly!"

We know the consequences. As we have seen, the personal computer has bloated into a mess that even its former masters, no mere "users," can no longer enjoy. Software logic tends to proliferate, features accumulate, and bottom-up means obscure big-picture ends. Even if systems were coded for elegance at every step of the way, the result would be programs, not experiences, and interfaces, not interactions. The idea of craft practice helps lead away from these conditions. Former Microsoft software pioneer Alan Cooper, who has become a leading voice of this approach, explained:

> We need a new class of professional interaction designers who design the way software behaves. Today, programmers consciously design the "code" inside programs but only inadvertently design the interaction with humans. They design what it does but not how it behaves, communicates, or informs. Conversely, interaction designers focus directly on the way users see and interact with software-based products. This craft of interaction design is new and unfamiliar to programmers, so—when they admit it at all—they let it in only after their programming is already completed. At that point, it is too late.[14]

As is commonly acknowledged, it is time for new practices in interaction design to use the prospects of ambient, haptic, and embedded interfaces as a way to reinvent computing. For this round of technological development to yield more satisfying results, its proponents need to account for abstract craft in the design and the use of pervasive computing. We need to operate between art and industry, work and play, and usefulness and beauty. We need to shape interaction design practices to be more in tune with the craft emphasis, professional judgment, and critical orientation typical of more established disciplines such as architecture.

The Need for Architecture

To interaction designers, the word *architecture* describes technological arrangements that, like buildings, are costly to build, are at least metaphorically scaled to be inhabited, are infrastructural in use, and are irreversible as results. Like buildings, computer networks in particular have been recognized as representations of their owners, major fixed assets, and agents of organizational change. As long as those networks were experienced through the desktop computer, they remained fairly separate from previous forms of architectural experience. But the more that they involve mobile and embedded systems, and the more that they develop unique aggregations in particular places, the more digital architectures become part of a physical structure. This is true both in terms of objects and experiences. When operations are collected in locally distinct, relatively persistent, and bodily memorable ways, the activity takes on aspects of architecture.

Architecture, in its very long history, was a social frame first and became operable equipment only later. (That is, architecture only recently acquired automation for the likes of heating and lighting; it always had some basic operations, such as opening and closing a door.) Computing has been the opposite. In its relatively short history, it was operable equipment first and social organization technology only later. As computing acquires ever more layers of software abstraction, however, such as local models of activity, it conceives schematic identities ever more independent of their technical execution. One word for these intentional identities is architecture.

The need to connect architecture and interaction design comes from overlapping subject matters and escalating social consequences. The two disciplines converge on the design of operable inhabitable systems. The path toward connection involves a shift from foreground objects to background experiences. The more recent discipline's path already follows that route. It is an evolution worth examining.

From Machine Interfaces to Architectural Situations

Interfaces

Interaction design tends to emphasize computer interfaces because those have been the first genuinely interactive technology. The prevailing computer-human interaction (CHI) model of interface design has been partly responsible for the current state of the desktop computer. The breakthrough on which the field emerged was the admission of psychological principles. The resulting graphical user interface has been the focus of the field of computer-human interaction for nearly 20 years. This interface is a virtual control panel whose design has remained quite technology-centered.

For almost as long, a keystroke-level model of goals, operators, methods, and selection rules (GOMS), has provided a quantifiable basis for usability inspection.[15] GOMS in its various forms is by now the most mature model of performance for graphical user interfaces. It remains especially practical for analyzing tasks within large software applications, such as CAD systems, in which there are many operators (i.e., lowest-level commands, or clickable tools and their modifiers) available for executing any given task.

A mechanical efficiency approach to interfaces has several advantages. It is quantifiable, which is required for legitimacy in the scientific community. It restrains programmers' usual tendencies to proliferate dialogues and options. It addresses the risks of physical injury from repetition. It becomes readily extensible into broader principles of usability, such as cognitive models of behavior and intent.

Usability metrics remain more an inspection and less an aid to conception, however. Thus their use perpetuates the notion of design as a corrective process. Also the metrics have little to say about off-screen context and how that shapes intent.

Meanwhile the interfaces that result shape our experience. As Jef Raskin has summarized: "Once the product's task is known, design the interface first; then implement to the interface design.... As far as the customer is concerned, the interface *is* the product."[16]

Usability for beginners may not produce the best experience for enthusiasts. For example, while a car with an automatic transmission is easier to drive at first, most people who find sport in driving prefer a car with a manual stick shift.[17] Much as one may recognize a good car less by its performance specifications than by what it is like to drive, so designers now hope to make the quality of interactivity memorable in itself.

Machines

It is telling that the largest membership organization of digital technology research professionals is named for machinery. This is historical; the Association for Computing Machinery has been around since computers filled rooms attended by specialists in lab coats. It is also metaphorical, because computers are not truly machines in the sense of powered mechanisms doing physical work. What it tells is that digital media began under industrial conditions in which the goal of technology design was mechanical efficiency with a minimum of human intervention. That intervention took the form of "operation," which tended to involve low-skilled attendance of a pushbutton control panel removed from generally large and dangerous "machinery." Following from the principles of scientific management, whose applications to design were best demonstrated in Taylor's time-and-motion studies, industrial engineers sought efficiency in the repetitive mechanical actions of using a technology. When today's software developers seek efficiency in the motor skills of moving a mouse about a screen, toward a goal of idiot-proof, button-based interfaces, they perpetuate this legacy.

Identity

Almost since its rise in the early twentieth century, the design of machines has been a question of corporate identity. To this way of thinking, design has less impact on how things actually work or get used and more on how they are known, purchased, and identified with. In a world where market share is more difficult to achieve than product functionality, this is of course important. Indeed, the economics of brand equity dominate much business theory today.

Because a brand is a relationship between a business and its customers, design identity is not necessarily confined to functions or appearances, but may be based on attitude. Some brands advertise lifestyles and not particular products at all.[18] Goods and services, which are easy to replace, are merely the tokens of this relationship, which is difficult to replace. Interactive media open new channels for establishing and building such relationships. Where the goods or services are digital, the branded experience is largely a consequence of interaction design. If the role of brand in other markets is any indicator, we should be on the verge of a tremendous diversification in interactive systems.

Questions of visual identity have informed much graphical interface design at a much more basic level of course. This has been understood mainly in terms of cognition. For example, the idea that a tool not only performs but also represents a process was important to the design of on-screen tool bars. Visual considerations prevailed; the "look and feel" of software interfaces, a popular way of referring to the quality of using them, was commonly observed to be all look and no feel.

The look has advanced enough for the data themselves to become subject to visual design. As Edward Tufte in particular demonstrated, visual identity is essential to communicating even the most quantitative data.[19] As Clement Mok observed, "Information design makes information understandable by giving it a context. Information design builds new relationships between thoughts and places."[20]

Now as computing becomes pervasive, the identity of these systems goes beyond the appearance of screens. New forms of ambient, haptic, and multiuser interfaces promote a shift from objects to experiences. Instead of emphasizing the visual identity of an object, under these circumstances we need to address the process of identifying with an experience. Although some people identify with using a system that is easy the first time, or in occasional use, others identify more with a system that rewards lifelong learning even if it is difficult at first. A violinist may identify with his or her instrument, for example. The identity of a design artiface is likely to involve an assertion about its usability in a context. This rightly discounts immediate novelty and first-time "idiot-proofing" in favor of usefulness and long-term assimilation. As the mainly mechanical business of counting clicks grows into a sociological study of usability in contexts, it should involves the psychological study of how people invest themselves in mastery of technology.

Context

By the late 1990s, ethnographic field work had become intrinsic to digital systems development. Contextual inquiry into the circumstances of technology use revealed social, organizational, and physical factors that influence the successful adoption of new technology as least as much as its functional features.[21] Work practices were the most usual focus of study, both because CHI had its origins in business technology and because in workplaces people are more disposed to tolerate new technology. Dot-commers were soon paying homage to people like the anthropologist Clifford Geertz, whose principle of "thick descriptions" became emblematic for how, because the observer influences the observed, contextual inquiry must set up the most appropriate rapport between the observing ethnographer-designer and the owner or client participating in a design.

In what became a standard textbook in interaction design, usability consultants Karen Holtzblatt and Hugh Beyer codified contextual inquiry.[22] They outlined a clear method by which software developers can gather field data on their customers before beginning work on technical solutions. The method begins from a master-apprentice relationship between the practitioners and their interviewers. (Here is an incidental defense of the craft aspect of high-tech work.) Context is "the first and most basic" of four principles for this contextual inquiry. "Go to where the work is to get the best data."[23] Sometimes the benefit is obvious, because there are a lot of mismatches in the corporate world. One telephone company had to actually see its line repairmen struggle with three-ring binders atop cherry pickers before they understood that they had to repackage their repair manual in a small spiral tablet that could be fastened to a worker's belt.[24] Sometimes the conditions are more subtle. Props and structures of ongoing work that appear to be incidental may turn out to be important to the appropriateness of additional technology. For example, the success of a change to portable e-mail devices may depend on whether coworkers can still leave really urgent messages on one anothers' chairs. Beyond its clear contribution to the ethnographic data-gathering techniques, then, Beyer and Holtzblatt's method moves toward a multivalent model of "seeing work."[25]

The more that factors external to computers per se become a design consideration, the more the design focus shifts from things to experiences (figure 7.2). The physical and social contexts extend the interpretation of the information context. Organizational, social, and physical factors play a greater role in usability. Incremental improvements in the efficiency of technology use may be undone or replaced by categorical improvements in the experience of technology use. Experience is of course in the minds of the users.

Activity	Design activities, not objects
Cognitive ergonomics	Minimize astonishment; maximize intuitive accessibility
Collective memory	Provide affordances for history; use enduringly legible elements; commemorate events; leave traces
Context	Expect physical location to provide protocols and constraints
Coordination	Versatility and satisfaction increase when actions involve tightly synchronized acts and multimodal reinforcement
Errors	Prevent errors; don't scold the people who make them
Flow	Satisfaction emerges when abilities are fully engaged toward objectives that are just about manageable
Latency	More satisfying designs tap latent ability
Scale	Images, objects, and actions have different meanings at different scales, especially relative to the body
Suspension of disbelief	Help people take part in representing shared objects and activities, but don't expect them to take that for reality
Tuning	Don't predict the state of complex systems; do let people customize, demonstrate, and accumulate the states of their technologies
Unintended consequences	Expect resources to be borrowed by insiders for unforeseen uses with discovered benefits, but also with revenge effects
Work practices	Tasks occur within a larger stream of conventions, the representation of which is essential to design

7.2 Common wisdom: a dozen axioms of interaction design practice

Users

Because factors of ability in context have become so important, observing and modeling users have now become essential to technology design.[26] The very word *users* unfortunately remains technology-centered, but at least we are being differentiated beyond the single, least-skilled "user."

Subjectivity is inherent to usability. Differences in abilities, intentions, and exploration processes affect the successful use of technology at least as much as technical features. One way to represent this is with a "cognitive walkthrough," which attempts to represent the sequence of assumptions, choices, and discoveries in the application of a technology. Interface designers of the GOMS school declare that they have been doing this all along, but they have generally done so at the level of individual sensorimotor operations. By contrast, this newer approach to user modeling has focused more on desire. Usability, identity, desire, and intent tend to relate.

User observation raises a classic problem of tacit knowledge. Much as pianist "falls out of the music" when asked to describe, or even just be conscious of, the actions of particular fingers, so other people cannot articulate their abilities even with less fluent processes. Minimizing the invasiveness of observation involves many sorts of trust and ethical agreements that give the inquiry a professional character. Contrasts between controlled laboratories and uncontrolled field conditions, between active interviews and passive monitoring, and between observing initial learning versus mastery all enrich interaction design. Moreover, in contrast to the more mature discipline of architecture, which is regularly accused of having forgotten its users (inhabitants) altogether, this newer discipline brings a more candid willingness to observe. Explaining technology use in context is a service.[27]

Of course, owners are more likely to adopt a design if they have had a voice in its inception and if their expectations have been given reasonable scope by first-hand encounters with the design problems.[28] Furthermore, when those owners are complex organizations or markets, design is more likely to prove appropriate if user involvement has occurred over a spectrum of participation that ranges from passive observation to active partnership. The practices of ethnography, contextual design, and participatory design all have legitimate roles in this.

In a common shorthand for representing users, designers often develop personae and scenarios. Personae represent differences in ability, intent, and context. Scenarios test designs with detailed stories of use.[29] Scenario planning explores alternative futures rather than alternative proposals. It uses sets of stories based on different outcomes to external questions.

Otherwise, user models are too often an attempt to fit preexisting market segments. In the lowest understanding of the word *scenario*, a single narrative describes how a user is unhappy in some regard until he or she acquires the product and experiences instant gratification. Where the technology is for production rather than consumption, too little is said about the experience, rather than the mere ownership—as if simply purchasing sufficiently usable gear will make everyone equally skilled. To study what makes us adopt technologies successfully is bound to make us to reflect on human nature in ways that help make design research into a liberal art. The impetus to buy may not be humankind's noblest aspect, however. Indeed, the psychology of induced consumption seems a subject that the liberal arts should take apart.

Experience

Recent design advocacy among the ranks of 1990s web luminaries has increasingly focused on "experience design."[30] This approach has provided a way to justify new practices in modeling and the economics of desire. It also extends the scope of design beyond individual technological systems to systems of systems.

Because a designed experience can alter a perception rather than accomplishing a task, the experience design movement emphasizes satisfaction. Satisfaction comes not just from meeting expectations, but also from changing them. Predictable formulas do not always produce satisfaction. Thus there is a paradox in the connotations of "experience design." Few of us want our experiences predigested. John Thackara voiced this concern quite well (figure 7.3):

I tend not to like or trust any all-encompassing experience that has been designed for me, and not with me: theme parks, shopping malls, air travel, most websites, 98 per cent of e-learning products. The majority of architects and designers still think it is their job to design the world from the outside, top-down. Designing in the world—real-time, real-world collaborative design—strikes many designers as being less cool, less fun, than the development of blue-sky concepts. To be fair, many younger designers feel free to set the stage for what is experienced. But the big money still goes to the control freaks. People do like to be stimulated, to have things proposed to them. Designers are great at this. But the line between propose and impose is a thin one. We need a balance.[31]

Fundamental errors still exist in the electronic media mogul's conception of experience design. If someone refers to interacting "with" an experience, then that experience has somehow still been conceived as a thing.

Big corporations often still approach design in terms of predictability. The experience they offer is guaranteed. The economics behind this was evident when Holiday Inn trademarked "No Surprises" 40 years ago. (Today, branded environments dominate the spaces of global travel. The global tourist industry tends to offer whatever the largest ranks of least-common-denominator travelers expect, even if that means golf courses and steak houses in Bali.)

When conducted according to behavioralist notions of inducing demand, "experience design" seems overly manipulative, and culturally sterilizing. But when allowing for unforeseen activities, this latest stage in the trajectory of human-computer interaction has high potential for cultural expression. Whether situated interactions develop a critical culture or are debased into standardized experiences depends to some degree on how well we can diversify design practice.

Articles of Association Between Design, Technology and The People Formerly Known As Users

Article 1   We cherish the fact that people are innately curious, playful, and creative. We therefore suspect that technology is not going to go away: it's too much fun.

Article 2   We will deliver value to people—not deliver people to systems. We will give priority to human agency, and will not treat humans as a factor in some bigger picture.

Article 3   We will not presume to design your experiences for you—but we will do so with you, if asked.

Article 4   We do not believe in idiot-proof technology—because we are not idiots, and neither are you. We will use language with care, and will search for less patronising words than user and consumer.

Article 5   We will focus on services, not on things. We will not flood the world with pointless devices.

Article 6   We believe that content is something you do—not something you are given.

Article 7   We will consider material end energy flows in all the systems we design. We will think about the consequences of technology before we act, not after.

Article 8   We will not pretend things are simple, when they are complex. We value the fact that by acting inside a system, you will probably improve it.

Article 9   We believe that place matters, and we will look after it.

Article 10   We believe that speed and time matter, too—but that sometimes you need more, and sometimes you need less. We will not fill up all time with content.

7.3 A widely admired manifesto for interaction design

Architecture

Appropriate design sets the stage for human experience. Like a great building, it reflects our aspirations, assists our daily rounds, carries collective memories, and provides a repository for many nonfiscal kinds of value.

We have seen how the fundamentals of activity theory point toward architecture, for instance. Embodiment is a property of interactions; persistent structures foster abilities and dispositions; social configurations give scale and type to fixed and institutionalized arrangements of built space. All of this becomes second nature; architecture is experienced habitually and in a state of distraction.

Architects' practices, which are far more seasoned than those of interaction designers, have been implicitly built around building types and place responses. Often an architecture firm becomes known for a particular *functional* type, such as libraries or laboratories. Often an architect's services involve exploring the appropriateness of different *architectural* types, such as courtyards and concourses, to a client's needs. Where an architecture firm concentrates on services, rather than signature works for patrons or delivering generic space as a commodity for developers, it normally operates in a particular *locality*.

Architects have long found design inspiration in new technology. Now they confront the addition of digital systems to the built environment. This shifts some emphasis from individual buildings to their interconnections, both physically as urban design, and digitally as networked cities. This brings the design focus to organizations and services. Teleserviced cities go beyond the still-consumerist orientation of experience design.[32] This expanded scope of architecture sets the stage not only for conterminous, synchronous activities, but also for the distributed, asynchronous flow of goods, services, organizations, and identities that have conventionally been the staple of design.

Ultimately there is something much more basic going on. Architects understand the importance of conceptual models. The plan precedes the building. Even at a mundane level, the organizational or social relationships designed are not the same as the structure built to support them. At a higher, more expressive level, some kind of conceptual premise shapes the eventual physical product. Architects know well how impossible it is to devote active, fresh consideration to every conceivable aspect of a building problem. Instead, a project concept gives emphasis to some particular set of considerations as the drivers of the process and the basis of schematic clarity. Such a premise rarely involves the accumulation of functional feature upon feature. Instead it revisits, and reawakens, some aspect of design solution that has fallen into thoughtless convention. As we have seen, this interplay of convention and reinvention gives vitality to enduring types and genres. This is one reason why architecture practices distinguish themselves by types. It also explains why the very notion of typology is an asset, not a liability to creative design. Interaction design thus reaches beyond interface mechanics and consumer experience delivery by means of critical reexamination of everyday life.

Toward a Critical Practice

A critical practice challenges prevailing values through works based in some other set of values. This is a form of conscientiousness. In a world where technique has too often become an end in itself, a culturally critical attitude has become essential to meaningful design. How to seek and identify a problem is as important as how to solve a problem.

The graphical user interface originally arose from this kind of questioning outlook. Now physical computing arises from questioning the assumptions by which the graphical interface became overblown. Projects in research institutes and schools often provide the best examples of this inquiry. Graduate students combine microelectronics with storyboards to demonstrate alternative formats of interactivity. For example, the first crop of projects from the new Interactivity Institute at Ivrea, Italy, in 2002 explored the emotional significance of smart objects based on interfaces other than screens and buttons (figure 7.4).

Critical practice in interaction design works with a more open-ended and provocative story than problem solving in device engineering has typically done. This emphasis on design as communication elevates our intentions about narrative structure, and leads directly toward assertions of a liberal art.[33]

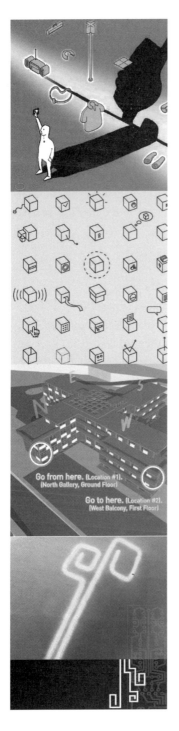

7.4 Design ideation: postcard images from set of research projects for critical design of mobile and embedded computing devices. (*Courtesy of Interaction Design Institute, Ivrea, Italy*)

A focus on premise has benefits in the practical work of building business applications. As John Carroll has observed, "The worst mis-step one can make in design is to solve the wrong problem."[34] Mere technical possibility is seldom sufficient in this regard. Rather, it is strategic insight that drives design. Problem seeking complements problem solving. Because problems are seldom determinate in any case, this seeking tends to involve selective attention to the more telling aspects of a situation. Designs become distinguished by which considerations have been given attention, among an excess of possibilities. Ethics may enter the decision making. Fostering a culture around design involves value judgments about what we want to build.

Typically, a design premise interprets a context. Good design helps us make sense of contexts.[35] The artifacts of design help us evoke understandings of contexts. Critical design produces artifacts that concretize and catalyze. Because cultural factors shape whether a design proposition raises questions of value, culturally situated propositions tend to yield more meaningful designs. A thing may be memorable more for when, where, or how it was acquired than for its intrinsic form. A service may be significant primarily in relationship to other services in time or place.

The success of a design is arrived at socially. Crudely, this suggests that market acceptance is the only criterion necessary. More to the point of critical practice, it suggests how design must help people understand a situation in a different way. The Walkman was a huge success because it made people rethink the context of listening to recorded music. Picture telephony has been unsuccessful for decades because people are unwilling to present themselves visually without advance notice. The amount of technology that has been adopted into everyday life demonstrates the significance of design as a successful liberal art. The very character of life and society has been transformed by planned artifice.

Appreciation and learned criticism must contribute to this social-ization and build a culture around interactive technology. Often these processes involve the maturation of genres. Much of the significance of productions tends to emerge relative to other productions. This is certainly true for individual designers, firms, and schools, whose authority rests largely in accumulated bodies of past work. It is also true for culture-wide bodies of work, however. Without going into the anthropology of expressive preferences or the reasons for the existence of genres, it is safe to say that cultures do become identified with particular themes. The Germans and French both bend metal, and both cook, but the one culture particularly excels at the former and the other at the latter.

Within genres, formal types emerge. In blues (a musical genre) the twelve-bar blues (a formal type) continues to yield more interesting new variants than switching to, say, eleven- or fourteen-bar blues. The ever-changing relationship between established forms and expressive content has long been at the center of aesthetic theory and appreciation.

Critical and cultural narratives remain essential to significant design, then. Good design is felt to be communicative. Arguments for design as a liberal art assert that it is principally a communication discipline. Arguments for the importance of artifacts assert that much of this communication is tacit. Cultural expression uses genres and their formal types as a means, not an end. Content is participatory; it is something you do, or perceive, and not simply information you receive.

When the objects of artifice pervade our lives, cultural narrative and memory depend on widening the appreciation of design. The true test of a medium is its capacity to support cultural expression.[36] Beyond usability and identity, we seek appreciation. Unless the kinds of deliberation generalized here can be built up around the interactive technology productions that increasingly occupy our efforts, those efforts are likely to result in cultural noise.

By limiting design consideration to that which is numerically predictable or visually fashionable, we produce a lot of junk. By expanding the design of context-based information technology to reflect appreciation, experience, usability, and desire, more of us can contribute to the cultural assimilation of so much technical production. We do not seem to mind being surrounded with books or buildings because those have been through much more such cultural deliberation. Depending on choices we now face in design practices, interactive systems could similarly assume cultural meaning. In any case, they seem destined to surround us.

8   Grounding Places

Life takes place. Our accumulated experience of intentional settings means a great deal to us, both as individuals and as societies. Design practices that foster this experience never go out of style. Perceptions of place may be subjective and fleeting, but grounding life in effective contexts remains absolutely necessary. Resorting to nostalgia hardly helps in doing this, however; there is little to be gained from understanding place mainly as something lost. At least to the more mobile and networked of us, place has become less about our origins on some singular piece of blood soil, and more about forming connections with the many sites in our lives. We belong to several places and communities, partially by degree, and in ways that are mediated. Common languages and technology standards have their advantages but also their disadvantages as mediators. With the rise of pervasive computing, more applications must enhance, and not undermine, our perceptions of grounding place.

At a cultural level, response to place remains an end in itself. After people themselves, places are the topic on which the greatest number of us have something to say. Any "sense of place" is notoriously difficult to study, however; it tends to be slippery, sentimental stuff. All the science in the world cannot explain what it is like to wake up in a great city. About all one can measure is the soaring cost of buying a bit of space there. We recognize that built places, especially great cities, remain essential repositories of value, whether fiscal, cultural, or personal. We value places enough that housing is usually a family's greatest living expense, tourism is the world's number one industry, and environmentalism has assumed aspects of a religion. Places are a way of taking part in the world, for with a resonance unequaled by many other aspects of existence, they are both socially constructed and personally perceived. Place imagery permeates language and legend, and place ownership dominates history and economics. Places are not just passive containers, but indeed the very expression of cultures. According to Yi-Fu Tuan, "Place and culture are interchangeable ways of looking at the same issues."[1]

Place and placelessness raise some of the largest philosophical issues of our age. Questions of global economics, industrial ecology, digital divides, and the disappearing boundary between nature and technology lie beyond the scope of almost any direct consideration, much less the present concern for architecture and interaction design. Meanwhile, the discourse in cultural geography reaches some of the most politicized, relativistic excesses of which the academy is capable. To review the literature of place and identity is to peer into a deliberately bottomless abyss.[2]

So this is not the moment to dwell on how telecommunication has annihilated distance, tourism has commodified culture, or global capital consistently tramples local patterns of value. There are few immediate solutions to the problem of how the building of environments, which is always a costly endeavor, becomes driven by the most shortsighted financial considerations.

Nevertheless we can, and must, temper universal information technology design with more helpful attitudes about place. The contextual design of information technologies must now reach beyond the scale of individual tasks to embrace architecture, urbanism, and cultural geography. No methodology exists for this difficult role, but this has already become a problem that is costly to ignore. Since place and culture are intertwined, it follows that more place-centered interaction design becomes a more culturally valuable endeavor. There is no escaping the fact that the world around us is being layered with digital systems. There is no denying our dismay at surveillance, saturation marketing, autonomous annoyances, and relentless entertainment. Whatever our desire for a "sense of place," we seem destined to get "places with sense." In more and more kinds of sites, the base background of our lives somehow becomes active. Smart spaces recognize at least something about what is going on in them, and then they respond.

Why Ground?

Digital ground is shorthand for a complex proposition: Interaction design must serve the basic human need for getting into place. Like architecture, and increasingly as a part of architecture, interaction design affects how each of us inhabits the physical world.

To engineers focused on means, this appeal toward ends may seem too philosophical. Yet philosophical questions seem more important than ever. Whether by design or default, technology has allowed a huge shift from a political attachment to a single home place toward a cultural connection with a multiplicity of nonhome places. Tribalism and nostalgia do little good in a networked world. A wholesome attitude to place has come to mean something else. More than ever, many places influence most lives, even the lives of people who do not move around. Now more of us are on the move than ever before. To the Earth's millions of migrants and refugees, life takes place in ways outside rooted origins. An unprecedented class of "high-tech nomads" (who are more likely the constituency of interaction design), call any number of places home. For us, technology provides not only mobility but also ways to connect. Because this activity is so often mediated, improved design can shape our desire and ability to connect with our surroundings.

Practical place-centered design must seek a middle way between a universal uniformity, which has been typical of high technology, and a local desire for completely belonging to one place, which has typically been antitechnology. (One can no more usefully be "against" technology than be against, say, cold rain.) Philosophically, this reflects a profound shift from using technology to overcome environmental limitations toward using it to understand and live more effectively within them. Interaction design must address how people move around, how they assimilate, and what kinds of local responses they encounter. As ever, design is for active, humane life; but without great precedent, now some contexts become active as well.

The word *ground* came from much earlier words for *bottom*. Its most usual meaning is some sort of foundation. In the graphic arts, for instance, the word applies to both the physical preparation and the communicative visual role of a base on which figurative or relief objects appear. In debate, and in theoretical derivations, ground describes an established theoretical base on which some current argument stands. In psychology, the word's connotation of firm support is important to any number of mental constructs. To be psychologically grounded is to have stability, resilience, and repose. In phenomenology, such a psychological sense emerges more distinctly where the ground has been modified. A horizon is more powerful in the presence of a church tower, for instance. A creative person can feel grounded in his or her work.

Embodied activity grounds satisfying interaction design. We have seen how persistent structures become a framework for the continued growth of abilities and understandings. By contrast, the absence of fixed contexts limits the development and flow of skills. With respect to physical technology itself, we have seen how embedded systems complement mobile devices, and how ad hoc encounters between the two raise the most interesting design problems. We have seen how local models build on protocols, and how architectural types embody established protocols as well.

All of these considerations shape a design pursuit of digital ground. Much as takeoff and landing are the most significant moments in flying, so connections and disconnections with other grounds highlight experiences of design (figure 8.1).

Infrastructures especially invite such design consideration. Like ground, "infra" means below; the two words share a basic meaning. Connections to local and regional infrastructures dominate our experience of technology. Among their more abstract consequences, these connections alter perceptions of insideness. Increasingly, the process is interactive. The highway tollbooth recognizes your car, the electricity grid tells you when demand is lower and power costs less, and websites research your browsing habits. One does not "interact" with an ordinary sidewalk, of course—one simply walks on it. Only when that surface deliberatively responds to you can the relationship be described as interaction. Only where multiple possibilities for interaction form an environment does this argument become relevant. And only in the context of place and community can we assemble these interactions into something we can understand as culture.

As in previous periods of history, many civilized souls can rightly deplore so-called progress; yet developments occur from which there is no prospect or desire for return, and being against technology per se is not a reasonable position. Is pervasive computing one of these developments? To continue the metaphor of takeoff and landing as the important moments in the experience of flying, this is tantamount to asking: "What happens when the ground comes up to meet you?"

8.1 The importance of contact: literally the most essential moment in a design experience, and figuratively a metaphor for the role of fixed context

Place and Space

Any philosophical agenda for situated design must compare place and space, place and community, or place and placelessness. Without going deeply into the dialectics of such comparisons, which fill volumes in their own right, it is worth briefly noting some distinctions. This should help clarify what place has come to mean and how our present concern reflects an intellectual watershed change.

We have seen how space has been fundamental to modernity. Architecture, the sciences, and global networking have produced space, but not necessarily place. These symbolic spaces have often overpowered the spaces of unmediated human experience. For example, David Hilbert's principle of representing any given problem by a multidimensional graph influenced cyberspace pioneers' claims that reality has moved into dimensions that can be perceived only by means of ordered symbols. Mathematicians indeed have far richer spatial constructs than architects, but these are only accessible through symbolic processing. Better interfaces can give form to these spatial abstractions, but all this capability has limits, which we have explored at length, especially in terms of embodiment in context.

Perhaps the simplest distinction between space and place was given by Yi-Fu Tuan: "Space is movement; place is rest."[3] Here are some other interpretations. Space is the anxiety of global indifference; place is the comfort of local malleability (another argument by Tuan). Space is alienation; place is identification (according to architectural phenomenologist Norberg-Schulz). Space is an ordering of understanding; place is an ordering of experience (urban planner Edward Relph, whose interpretations of place and placelessness were among the best of the especially profuse work on this topic in the 1970s). Space is a social production; place is a personal reading (Henri Lefebrve, who reopened the study of emergent social spaces). Spaces are the basic divisions of our surroundings; place is our history and adaptation of them (landscape historian J. B. Jackson, who legitmized vernacular landscape studies). Space is the scene of being; place is a site where human modes of being are well provided for (Heidegger).

Intellectual reactions to the intrinsic excesses and shortcomings of modern space have surfaced in many disciplines. In one particularly authoritative long-term history of the topic, Edward Casey has observed that although place has been "dormant" in western thought for hundreds of years, now it has returned to respectability.[4] Space has been "supreme." Whether by scientific, theological, or architectural construction, modern space has been abstract, absolute, extensible, metric, universalizing, objectifying. If it has not been lost outright amid these spatial tenets so central to modernity, place has at the very least been reduced to mere location. The fate of place, according to Casey, is to reemerge not as a "total phenomenon" in direct contrast to the "monolith" of space, as it was in its ancient philosophical foundations, but as a multitude of phenomena overlooked by modernity. Embodiment has been one such necessary but unquantifiable phenomenon that has regained academic stature.[5] Casey's literature review illustrated a much more broadly based interest:

Common to all these rediscoverers of the importance of place is a conviction that place itself is no fixed thing; it has no steadfast essence. Where Heidegger still sought something resembling essential traits of place (e.g., gathering, nearing, regioning, thinging), none of the authors I have just named is tempted to undertake anything like a definitive, much less eidetic [visually exact], search for the formal structure of place. Instead, each tries to find a place *at work,* part of something ongoing and dynamic, an ingredient *in something else*: in the course of history (Braudel, Foucault), in the natural world (Berry, Snyder), in the political realm (Nancy, Lefebrve), in gender relations and sexual difference (Irigaray), in the realms of the poetic imagination (Bachelard, Otto), in geographic experience and reality (Foucault, Tuan, Soja, Relph, Entrekin), in the sociology of the polis and the city (Benjamin, Arendt, Walter), in nomadism (Deleuze and Guattari), in architecture (Derrida, Eisenman, Tschumi), in religion (Irigaray, Nancy). To read the bare list of names is to become aware of a far-flung and loosely-knit family resemblance of changing and contingent traits.[6]

As the study of persistently emergent perceptions, phenomenology reopened the way for academic work on place; for the more that modern man could shape his environment at will, the more he correspondingly lost touch with it. Phenomenology, particularly in its formative expressions by Heidegger, reexpressed how the human condition involves the right balance of divine aspiration and mortal earthliness. (As evidence that a high-tech building can achieve this, Santiago Calatrava's Quadracci pavilion is both sensate and sensuous; figure 8.2)

As one essential component in Heidegger's famous fourfold construct, which the act of building "gathers,"[7] the earth serves as supporting resource (in effect, ground) to be inscribed on in turn by human infrastructures. Heidegger foresaw (and notoriously fell into, for which he is now regularly dismissed *ad hominem*) the political dangers of retroromanticism, overattachment to place, and all the other aspects of what we now experience as the violent tribalistic backlash to globalization.

8.2 Sensate form: Milwaukee Art Museum Quadracci pavilion designed by Santiago Calatrava. (*Photo by Jim Brozek, courtesy of the Milwaukee Art Museum.*)

Nevertheless it is this phenomenological thread in twentieth-century thought that led to the interpretations of embodiment (Merleau-Ponty), contextual perception (Gibson), and situated action (Suchman) that ground current work in interactivity. When interpreted at the social scale, this history also led to urban typology (Rossi), inhabitation patterns (Alexander), and place identity (Norberg-Schulz). To the latter, "the environmental crisis is evident [even] in the flat neutrality of a domestic interior wall."[8] This kind of phenomenology explained the weaknesses of a world become all space and no place. "Most modern buildings exist in a 'nowhere;' they are not related to a landscape and not to a coherent, urban whole, but live their abstract life in a kind of mathematical-technological space which hardly distinguishes up from down."[9] In short, the present inquiry into contextual computing belongs among countless examples of place reemerging in recent thought.

Too often place exists mainly as a sense of loss. Who among us has not lamented the ruination of a favorite spot? Lately too much can happen anywhere; too much is the same wherever you go; and too much has changed the next time you come back. Like bad weather, creeping placelessness makes excellent small talk because it afflicts everyone and yet none of us can do much about it.[10]

J. B. Jackson marveled: "I am bewildered by our casual use of space: churches used as discotheques, dwellings used as churches, downtown streets used for jogging, empty lots in crowded cities, industrial plants in the open country, cemeteries used for archery practice, Easter sunrise services in a football stadium."[11] Meanwhile, as explained by the media historian Joshua Meyrowitz, electronic connections have eroded our ability to play different roles onstage and backstage in our lives.[12] When we never know who is watching, or conversely, when we can watch activities at social remove without having to make corresponding physical passages, then we tend toward fewer distinctions among places, and ultimately toward some tyranny of the casual. Instead of having to go to the palace gates to get a glimpse of the king, we can stay home and see television footage of the president out jogging.[13]

Disembodiment accelerates as electronic representation of the city engenders a receptivity to the virtual. As the urban historian M. Christine Boyer has explained, this occurs in several ways. First, people retreat into private gated zones with their media, where the "clean" computer contrasts with public squalor, and so they spend less time in the physical places of the city. Second, the imaging of the city resorts to increasingly commercial and privatizing strategies to lure people back outside from their screens. As predictable trademarked formulas are applied, brands become places and places become brands. Third, the spread of these nonpublic urban spaces subverts any preexisting legibility—or mental mappability of the city. "We have seen how throughout history the body has been projected onto the image of the city, and how the city has been described as a simulacrum of the body. As body awareness withers, space becomes immaterial; as we retreat into the privacy of our media-altered realms, the direct experience of the city disappears. We no longer read the city as a totality."[14] This dissipation costs dearly because the city had been the main cultural memory device. The city was a theatrical panorama of forms commemorating, officially or unofficially, its inhabitants' history and aspirations.[15]

In a world of universal experiences we are thus, as the architectural historian Anthony Vidler put it, "always and never at home."[16] This has produced a loss at the level of embodied predispositions, habitual contexts, and cognitive accessibility. "In this city, where suburb, strip, and urban center have merged indistinguishably into a series of states of mind and which is marked by no systematic map that might be carried in the memory, we wander, like Freud in Genoa, surprised but not shocked by the continuous repetition of the same, the continuous movement across already vanished thresholds that leave only traces of their former status as places."[17]

Placelessness has its obvious worldly sources. You know these complaints. People move around much more, and all this motion accelerates change, increases chaos, undermines local differences, and decreases the importance of meaningful dwelling at home. Meanwhile electronic media eliminate distance and differences, not only across physical geography, but also across social hierarchy. Architectural and urban spaces are no longer a match for mathematical space in quality and sophistication; cyberspace was allegedly created by the takeover of the former by the latter. Disembodiment by electronic media undermines the anthropomorphic qualities of the city; as noted earlier, the "body politic" becomes less legible. As evident in a demand for constant entertainment, first-hand experience seems to be in short supply; virtual passes for reality. An automobile monoculture breeds disembodiment as well; some chauffeured suburban children have never gone for a walk. The use of space is highly segregated, not only by function as a consequence of outdated industrial zoning codes, but also by economic and social niches as a consequence of market stratification. Advertising rules; more space is designed to grab the attention of casual, passing users than to reward subtler appreciation by regular inhabitants. Material overproduction delivers increasingly interchangeable goods and services in increasingly uniform and gigantic settings—the big box stores. Shortened fiscal horizons lead to impermanence in building and to transience in inhabitation. Relentlessly casual use of space, amid a confusion of scales and a dominance of images, accelerates the cultural process the poststructuralists described as "dislocation." Sooner or later, just about everything appears out of context. It is all a familiar rant.

The only way to treat such symptoms is to change their basic cause. However powerful these many manifestations may seem, this distress is more fundamentally a consequence of a modern philosophy in which place scarcely bears mention, but space and time have become primary constructors of experience. Everything points to questions and reconsiderations of value—but not necessarily to relinquishing the quality of life. Almost every philosopher who has explored such reconsideration has advocated a greater role for design. To arrive at new outlooks on value, however, as the present inquiry aims to do in the end, we must first understand the role of context and place in technology, and then see how places emerge around nodes among technologies.

Place and Community

Once there was less ambiguity about place. Either you were inside the city walls, or you were not. If you broke the rules on the inside you might get "rusticated," that is, thrown outside to fend for yourself. Today the question of belonging drives much more complex politics of identity. This is the mainstay of the academic culture wars, in which "alterity," or otherness, is nearly mandatory. Defiant young scholars readily assume conspiracy and alienation. Someone else is in charge of a zero-sum game in which one's own group is marginalized and victimized. Dutiful poststructuralists make texts of all manner of tacit social divides across previously undertheorized boundaries. However inaccessible it has been made, this social criticism has revealed two significant directions away from politically violent notions of place. First, the notion of place does not have to consist of a nostalgia for origins, and second, the perception of place does not consist solely of discovery but involves active construction of inside-ness and outsideness.[18]

Normally a place reflects a tradition of appropriations. As Edward Relph put it, "Places are defined less by unique locations, landscape, and communities than by the focusing of experiences and intention onto particular settings."[19] Thus while we can speak of the identity *of* a place, we must also admit identification *with* a place. Place is as much about subjective insideness as objective boundaries. Physical boundaries may just as easily be the cause or the effect of social and cultural memberships. Space lies outside the walls, or outside the social sphere, but the experiences of place occur inside these seen and unseen boundaries. Insideness varies by intent, by belonging (e.g., insiders), and by insights into the nonstereotypical aspects of place. Outsideness may take the form of personal alienation, cultural difference, or economic formulas in which the operands no longer have referents; or it may simply be the indifference of a worker to whom the present setting is of little importance.[20] Once there was much less ambiguity about belonging.

To Relph as to others, it was possible to describe some degree of grounding in terms of authenticity.

> "Inauthentic attitudes to place may be unselfconscious, stemming from an uncritical acceptance of mass values (kitsch); or they may be self-conscious and based on a formal espousal of objectivist techniques aimed at achieving efficiency....
>
> An inauthentic attitude to place is nowhere more clearly expressed than in tourism, for in tourism individual and authentic judgement about places is nearly always subsumed to expert, or socially accepted opinion, or the act and the means of tourism become more important than the places visited."[21]

When insideness is superficially engineered, then damage occurs. When insideness involves physical co-location but not full literacy in local signs and symbols, much less histories, then disruption occurs. The work of "uncommitted insiders" has been the strongest source of placelessness.[22]

All this of suggests aspects of community. Few words have been so badly abused. "Community" has come to mean any aggregation of people, whether by shared interests, shared market behavior, or proximal location. In better usage, the word describes active participation based on a sharing of goals. With respect to location, it means not only a co-presence, but also an interweaving into some dense web of social relations. This meaning of community became quite topical with the onset of online socialization, but it remains more applicable to everyday life in physical neighborhoods. Everyone makes some kind of rounds through a set of support services in convenient proximity. The rounds are made using various technologies, however, and proximity is not necessarily physical. Habitual, participatory appreciation of local resources creates a feeling that only this place uniquely fulfills one's complex set of needs.

Place and community differ in that the latter is often distributed and without physical features. Place and community are similar in that each involves a perception of insideness. In the case of community, the perception is more social. A location may be very much a place to one person and not at all to another. For community, however, there must be some commonly held perception, not only of its existence, but also of its purpose. This is a prime example of how contexts affect intent and institutions confer identity. There are countless ways to describe the perception of community, and the literature built from the network technologists' studies alone is extensive. One such set of community attributes follows.[23]

Trust. There remains no better basis of community than a cooperative endeavor toward common goals. Thus the belief that goals are held in common becomes fundamental. Where co-location no longer remains necessary for participation, online design becomes about building trust and collaboration.

Support networks. A community provides a dense web of services for getting things done. These transactions may be commercial or noncommercial, and their objectives may be economic, civic, or personal. Often this web is situated; consider how a neighborhood consists of the right mix of services.

Common experiences. The experiences that people have been through together bind them as a community. They create a common cognitive background and disposition. Where the experiences are voluntary, they reflect the community's goals. Where they are incidental (e.g., dealing with a fire) they reflect its cooperation.

Interests. The "lifestyle clustering" so prevalent in today's self-fulfilling geodemographics is a community in itself—communities are not mere market segments—but it becomes an inevitable component of community. Shared interests do correlate with trust, services, and experiences.

Type of transactions. The way in which people deal with one another and the kinds of exchanges that are likely characterize a community. Where transactions are restricted to buying and selling, or to agreement over consumer choices, this character is correspondingly limited. Where unpredictable transactions are welcome, trust exists. Invented transactions, as in art works and games, can form impromptu communities (figure 8.3). The one best indicator of a larger community is a willingness toward casual social encounters with persons having different goals, experiences, and interests. One good word for all this is civility.

Enforcement. Trust in interactions depends on the exclusion of antisocial activities. Every community must have insiders and outsiders, and some means of maintaining that distinction. A community is defined by what it considers antisocial behavior. These restrictions may be tacit. A good mark of community is where "there are no rules, but don't you break any of them."

External adversity. The bond of common experience combines with the need for insiders and outsiders to make external adversity a strong shaper of community. Friendships are made amid hurricanes and wars.

History. The remembrance of adversity, or of particular qualities of civility, becomes a basis of social identity. This is true whether or not these memories are of direct experience; every community has its myths. One may belong to a community in part just by ascribing to its histories.

Landmarks. City and landscape are commonly regarded as memory devices. History tends to be linked to geography. Built landmarks

8.3 Location-based play for insiders: in an early example of location-based wireless play, based on the cells of the telephone grid, this wandering cat showed up on participants' screens at changing locations around London. Fiona Raby and Ben Hooker, Royal College of Art. (*Courtesy of Fiona Raby.*)

symbolize founding principles, civic processes, and expectations about the nature of social interaction.

Proximity. All of the above dimensions interelate. With enough of them in force, proximity need not be an attribute for a strong sense of community to exist. Proximity alone, as marketed by the builders of gated residential subdivisions, for example, will hardly suffice. Although it is no longer necessary or sufficient, physical co-location nevertheless remains a likely condition for many of the dimensions of community given here.

High-Tech Nomads

An unprecedented mobility transforms our usual notions of belonging. At least some of the world's migrating millions travel by choice, with ample means, and often. Their lives unfold on several continents with scarcely any knowledge of their next-door neighbors on any of them.[24] American-educated, Indian-born engineers meet over Vietnamese noodles in Amsterdam. Enterprising twenty-somethings in Seoul and São Paulo have more in common with one another than they do with their respective parents. Like changing languages in mid-sentence, they switch deconstructively between world views at will.

Lives themselves are the places these people know, and the kaleidoscope of cities and countries around them just keeps turning.

The pile of devices in one's carry-on keeps growing. For example in 1999, iGo.com, a company that trademarked itself a "mobile technology outfitter," had 6500 mobile products and accessories in its database.[25] Here the weary air traveler could upgrade from earplugs to noise-canceling headphones. He or she could order up a *Quicktionary* hand-held optical text scanner for translating printed words from Spanish, French, German, or Italian into English. The hotel user tired of failed hookups could purchase a "Road Warrior" connection kit that included phone and printer cables, couplers, adapters, a polarity reverser, and a line tester. A behind-the-wheel business person could order a suite of mounts so that his or her laptop, PDA (now fitted into the cup holder), voice recorder, and cell phone would no longer be thrown to the floor on sudden stops. To this, an orienteering sort could add a satellite-to-dashboard GPS navigation system showing real-time updates of one's own position on street maps, together with expansion cartridges for finding decent restaurants.

Service Ecologies

Being on the inside means getting better services. This is as true within corporations as within cities, and as true of the global polyglot metropolis as it was for the ancient *polis*. Insideness consists of knowing where to obtain what one needs, in what arrangements of urban space and time schedule, and now, increasingly, by what degree of technological mediation. In a truly great place, each of us cultivates our own carefully refined set of regular haunts. In a lesser place, more people have to live more similar, less imaginative lives, on the limited basis of available services. In a mobile world, different places offer different qualities in the essential experience of making the rounds.

Clusters remain essential to this activity. The right combination of services characterizes a neighborhood (figure 8.4). The convenient proximity of complementary activities, say a bakery and cleaners, matters there. A finer differentiation within a particular class of services

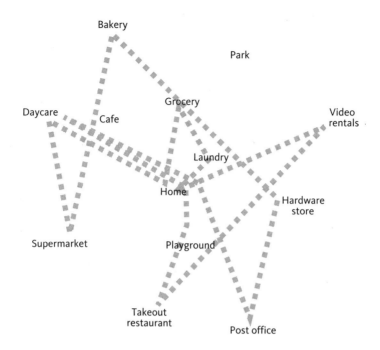

8.4 Making the rounds: service ecology in a neighborhood routine

characterizes a district. Interconnections of various resource flows matters more in those cases. The interrelationship of different scales matters in the form of neighborhoods, in the flow of goods and services, in the radii of the markets served, and in the patterns of inhabitation. Localized adaptation and evolution are important to vitality. The best clusters are never formulaic, the best (and worst) things that happen to a neighborhood tend to be incremental, such as the opening or closing of just the right café.

The density and diversity that we now know as the basis of good urbanism were the very targets of modernizing "urban renewal." Jane Jacobs' epoch-making attack on the planning establishment in the name of "city life" still holds lessons. "Intricate minglings of different uses in cities are not a form of chaos. On the contrary, they represent a complex and highly developed form of order."[26]

Systems thinkers recognize aspects of ecology in the vitality of good service networks. When more activity is devoted to upkeep than to expansion, when elements at very different scales depend on one another, when elements co-evolve on the basis of context, and when resilience operates at the level of whole systems, then some sort of ecology is at work. A service ecology is a self-regulating aggregation around some sort of flow, which yields some sort of systemic advantage. This idea is quite familiar to today's proponents of "just-in-time" economics and supply chain management, whose goals include economic stability, resource economy, physical identity, and long-term adapability. Not just incidentally, these are also qualities that we admire in neighborhoods or regions.

The elements of service ecologies become increasingly subject to electronic mediation. Remote connections complement face-to-face meetings, both in space and time. Spontaneous gatherings of individuals, communities, or businesses become easier with mobile communications, especially when they are supplemented by positional information or proximity polling. The latter means asking who else is around, a fundamental human query, which lately has been carried over to online chat by ICQ, and now back to the physical realm by positioning and short-range message systems often built into the latest generation of mobile telephones. Digital networking serves physical clustering in other ways as well. Referral networks, so essential to clustering, become faster and better filtered. Small buyers and sellers formerly without means of finding one another are brought together through the net. Documentation feeds, such as those from sales transactions, provide terabytes of data to be analyzed for systemic patterns. Picks and finds, specifically situated within the rich chaos of the city, get shared and discussed in new media.

With respect to agendas of interaction design, the notion of ecologies translates from service ecologies into design problems at four fundamental levels. To begin, *device* ecologies emerge within the ad hoc encounters of mobile and portable technologies in contexts. Here the principle applies that optimizing the parts may disoptimize the whole. Feature-focused engineers must turn their attention to how devices interoperate and evolve into roles. Attention to systemic flows (again, whether of resources, goods, services, or communication) becomes the subject of device design. The principles of industrial ecology apply.

Next, and at a level without which such hardware interoperability cannot occur, designers increasingly recognize *information* ecologies. According to Bonnie Nardi, who has done much to popularize the term, information ecologies manage knowledge by a combination of software models, contextual configurations, and human reflection in action.[27]

A natural environment offers many toeholds for life of various forms. With tenacity and vigor, species migrate and change to fill the available niches. These adaptations lead in turn to further change, as the entire system adjusts to new constraints and possibilities. A healthy ecology is not static. The pace of new technology development ensures that school, work, and home settings will continue to be offered newer, faster, and different tools and services—not just once, but repeatedly. Evolution implies a past as well as a future. An information ecology acquires its own history. It reflects the stable participation of an interconnecting group of people. An individual session with an automatic teller machine, for example, is not an information ecology. It is a useful but isolated service. By contrast, a bank office with diverse services, tools, activities, and interconnections among people is easier to interpret as an information ecology. When we are in a bank, we can sense that the technology arrangement has a continuing history of development and change.[28]

Third, such situational types increasingly become subject to the design of *interaction* ecologies. Diversification in interfaces makes as much sense as diversification in the urban form. Clusters of experiences of mediated interactions make as much sense as clusters of previous services and communications.

Ultimately, these agendas become those of *architecture*. In setting the stage for habitual activity, in representing organizations to the constituents, in creating enduring structures that reflect and perpetuate particular attitudes about interaction, designed service ecologies are architecture. When the technological means increasingly embed physical computing into elements of buildings, these newer subjects of design become part of what we already know as architecture.

Technologies of assimilation have long existed in the form of signage, guidebooks, and protocol manuals. Now by means of tags, sensor fields, positioning systems, and the many other situated technologies explored here, these instructions become more subtle, more diverse, and more available. For example a simple tourist guidebook is replaced by position-indexed databases which can provide a great deal more, and better filtered information about where one is without the cost of carrying about information on where one is not. Social navigation systems such as quick messaging or local polling on handheld devices add a dynamism to the act of being out on the town, and they do so in a way that adds possibilities and plays to the meaning of insideness. Distributed monitoring systems can interpret patterns, whether of current activities, long-term use, security, pollution, or much else in ways simply not observable before. Local business sourcing connects smaller suppliers both to global networks of demand and to related suppliers of necessary supports. For example, a small winemaker in Tuscany can join with other winemakers elsewhere for economies of scale in distribution and publicity, and at the same time can identify local sources of necessary materials.[29] Bottom-up networks of support services can index themselves by physical location. New arrivals to a neighborhood can find their way around the place by means of such referral networks. Supply chain managers of global enterprises can seek providers in a manner more in harmony with local service ecologies.

Getting into Place:
Architecture, Interaction, and Ground

The cultural geography raised here may seem like a digression, but it is not. Meeting the design challenge of pervasive computing has to involve some larger, unquantifiable pictures of place and community. Supporting the way we belong to multiple places and communities has to improve our dialogues on what those places and communities are. A place is not just some positional coordinates. Community is not just a marketer's mailing list. Rather these are complex, subjective perceptions in which the nature of mediated interactions plays a vital role.

When experience flows we get into place. Flow is of course an essential goal of interaction design, and fixity is an essential goal of architecture. Now the two join. To complement the spaces of information with the contexts for getting into place, it helps to think in terms of ground.

9   Accumulating Value

Places interest many of us, but what makes them worthwhile? Why exactly are reports on the irrelevance of geography so mistaken? What value does context create? For a theory of place to become a practical agenda in interactive technology, it must articulate its economic value. The rise of pervasive computing depends on better valuation of context for its success.

Places are essentially repositories of value, although not necessarily of measurable, convertible, instrumental value (figure 9.1). Indeed, places are accumulations of capital, whose increase needs more accurate economic representation. Depleting that kind of capital cannot be treated as free goods or net income. Correcting this accounting error has become of the most fundamental goals of practical environmentalism.

Like cultural geography, environmental economics has become significant to more disciplines than just those dealing with natural resources. It informs how the new field of interaction design now moves beyond device engineering and into the architecture of places. When location matters to the appropriateness and therefore to the success of a technology design, then the contribution of that design to the worthiness of that location becomes important. When the evaluation of such worth is in play thanks to new kinds of interdisciplinary collaborations and new emphases on adding value through strategic design, then a review of first principles is warranted as well. It is worth repeating that for any genuinely new economy, the definitions of value must change.

9.1 Three forms of accumulation: buildings, cash, and technology. (With apologies to, and after the well-known drawing by, Leon Krier.)

Value Emerges from Interactions

This expansion of intellectual context can be justified partly by its simple point of departure: interactions establish value. Value emerges from interactions. This inquiring and examining has no end: what matters to individuals, societies, and markets never reaches a final equilibrium, but remains constantly in play. Value play unfolds daily—from the routine of domestic micro-transactions such as buying a quart of milk to the enormous throughput of global capital markets. Value is deliberated at all levels from the highest courts of law to the lowest manners on the street. Child rearing and education instill moral value; cultural expression haggles over aesthetic value; and frenzied trading demonstrates the highly interactive, ever-fluctuating nature of economic value.

These value deliberations literally take place. Each commodity has its market; each art has its stage; and each morality has its church. Traditionally, such social infrastructures took physical form as landmarks and districts—public space to punctuate the otherwise private continuum of the city.

Mentally mapping these places contributed to individuals' cognitive background on what values were at issue in the society. Habitual contexts institutionalize particular forms of deliberation. Legible environments confer identity.

As receptacles and channels of value, the institutionalized sites of deliberation become natural candidates for design. Thus public places have traditionally been the focus of architecture, and thus now digitally augmented places become a focus for interaction design. Again, markets, and by extension, electronic commerce, are just one example of this. When portable and embedded information technologies augment a full range of built environmental types, approaching those in terms of interactions becomes a much greater value proposition.

Like anything else today, design must be economically justifiable. It may also expand what is considered viable. Can good design remind people that buying and selling are just one example of deliberating about value? Can place-centered design demonstrate how value transcends price? Or might markets become the only locus of interactions, and economy the only remaining form of value?[1]

When "adding value" is the ultimate measure of legitimacy, few topics receive more attention than what constitutes value itself. Type
"theory of value" into an Internet search engine and you will retrieve
thousands of items in response. Most of the retrieved items will be
business texts (often mission statements), however; and most of the
value being added is economic value. This is because economic value
can be more readily agreed upon than, say, moral or aesthetic value.
There are many kinds of value, then, and some prove more measurable than others.

Because value relates closely to such essential abstractions as
desire, utility, cause, and good, any attempts to address it quickly
become philosophical. Philosophers have perennially asserted value as
a fundamental category of thought. Thus a bit of philosophy is warranted for present purposes. To ask how design might help normative
conceptions of value evolve, it will help to have some idea of what
value has meant.

Value requires philosophy because it remains inescapably subjective. As indicated earlier, the traditional positions in value theory
maintain that nothing has value in itself—it has value only *to someone*. However, modern, objective positions counter that value can
exist independently of human intent. Are the things we desire in life
good because they are preferred, the economic philosophers ask, or
are they preferred because they are good?

The question of value not only swings between subjectivity and
objectivity, but also across many other classic axes of inquiry.
Consider the distinction between instrumental and intrinsic value, for
instance. Instrumental value concerns what something is for, but
intrinsic value concerns what something is. The former is evident in
the fact that many economic standards define value in terms of utility.
The latter can be explained by a standard ethical trope: You may kick
a stone along a road, but you shouldn't kick a mouse along a road. As
a living entity, an animal has intrinsic value.[2]

More subtly, note how instrumental value also relates to aesthetic value. In approaching the latter, one cannot ask what a work of art

is "for." Art need not be instrumental. One may ask what a work of art "says", however; and one may believe that an aesthetic production that says nothing to anyone has relatively little value just for existing. Aesthetic value must be culturally situated. It exists mainly at the convergence of qualified opinion. This may be what makes aesthetic value suspect to scientists: It is neither apparent nor consistent to everyone. Aesthetic value needs theory—and therefore critics—by which to deliberate its subjective expression and interpretation.[3] These in turn benefit from being grounded in objective constructions, such as tonal scales in music, and genres, such as portraiture in painting.

Intrinsic value needs no theory beyond the assumption that life is good. "There is no wealth but life," declaimed John Ruskin, one of the most vociferous critics of early industrial materialism.[4] If for altogether more sophisticated reasons, nevertheless many twenty-first-century thinkers seem inclined to agree. To those of the new school of genetics and complexity theory, value is intensive rather than extensive. On the whole, value is found in the increase and involution of life, and not, as Ruskin had criticized the industrialists, in the "getting on" to more things and more places like the ones we already have.[5]

Ruskin's view is an example of moralizing. Moral value, perhaps the most difficult value form of all, provides a means, for example, to uphold cultural identity and human rights. Its application typically serves the sanctity of living things, particular human cultures among them, as its ends.

Philosophers have thus often explained value in terms of teleology, the study of purposes. Whereas a means causes something to come about, and may be valued for how conveniently it does so, an end is something to be pursued, and may be valued for the effects of the pursuit as well as the attainment of the goal. The latter is evident in the truism that the things really worth doing in life take practice. Then of course, the ends of one particular activity may be valued as the means toward another. This leads to a chain of activities, which philosophy hopes must lead somewhere. That goal serves as a final cause. Moral value deliberates about final causes: individual versus societal rights, duty versus pleasure, worldliness versus divinity, etc. Moral value reminds us that the ends do not necessarily justify the means.

To the more abstractly minded, subjective questions of valuation thus link to objective questions of causation. Note that before Newtonian physics, causation was interpreted more generally as an explanation, and less specifically as an ordering of events. We do not think of bronze as the cause of a statue, but just such an instance appears in Aristotle's *Physics*. In that treatise the "four causes" were *causa materialis, causa efficiens, causa formalis,* and *causa finalis.* Each of these provides a basis for understanding value added, and one of them, the *causa formalis,* clearly indicates the importance of design.

Consider the standard example of building a house. The material cause (i.e., explanation) is the harvest of wood and stone from which the house is built. This has value for its availability and malleability. The efficient cause is the work and tools by which the house is built. This most closely approximates modern, labor-oriented notions of added value. The formal cause is the organizational concept according to which the house is built. (Note how this cause corresponds to postindustrial notions of value added through reorganization.) The final cause is shelter. Internally, each of these causes has its own means and ends; and externally, the final cause here may be a formal or efficient cause toward some larger cultural goal. The purpose of the house is shelter, but the purpose of shelter may be bodily fitness to take part in civilization, the ends of which may of course be the glory of God, the might of the state, the accumulation of wealth, or whatever else constitutes final value.

The notion of final value addresses the aspirations and root causes of existence. This notion's very existence reminds us of the absurdity of a society in which efficiency is all that matters. Instrumentality makes no sense as an end in itself. Yet we live, work, and design in an age of means. As if in some perversion of the uncertainty principle, the more we know how to do things, the less we seem to know what to do. This is a classic problem of means and ends.

Pursuit of ends by any means can be especially dangerous, as the totalitarian regimes of the twentieth century so thoroughly demonstrated. But how well do most people understand that our emerging twenty-first century's apparent pursuit of means toward any ends may be little better? To quote Ruskin again, "All healthy people like their dinners, but dinner is not the main object of their lives. So all health-ily-minded people like making money—ought to like it, and to enjoy the sensation of winning it; but the main object of their life is not money; it is something better than money."[6]

Much as means alone make little more sense than ends alone, efficiency alone makes little more sense than aspiration alone. From the more timeless perspective of Aristotle's categories, one may say that modern economics has dwelt disproportionately on the *causa efficiens*. The *causa materialis* it has taken for granted as unlimited resource income. The *causa formalis* it has addressed in part, but generally as conflated it with the *efficiens*—less as design and more as management theory. And as for the *causa finalis*, modernity appears to have accepted economic activity as an end in itself. "Adding value" is justified in any form that is measurable and monetized.[7] In this spreadsheet view of the world, the measurability of increase at least partly justifies the means. Here then is one predicament of value. Outside of such demonstrable fiscal gain, how well can the mechanisms of modern economics ask what an economic activity is "for"?

Utilitarian Value

It is possible that an economic activity can be "for" enabling another economic activity. Thus you make money by helping other people make money.[8] How woefully impractical is any alternative. As evidence for ours being an age of practical means, consider that many standard references on economic philosophy refer their treatment of value to an entry on utility.[9] This shifts the problem away from moral value toward economic value. Utility has long been the route that political philosophers have taken toward any more comprehensive theory of economic value.

In the practical philosophy of John Dewey, for instance, the theory of valuation revolves around the "ends-in-view":

> It is plain as can be that desires only arise when "there is some-thing the matter", when there is some "trouble" in the existing situation. When analyzed, this "something the matter," is found

to spring from the fact that there is something lacking, wanting, in the existing situation as it stands, an absence which produces conflict in the elements that do exist. When things are going completely smoothly, desires do not arise, and there is no occasion to project ends-in view.[10]

According to Dewey, value does *not* arise when activity is instinctive, organic, or habitual. (So much for the intrinsic value of living patterns.) Only when a transforming desire and its object emerge from settled conditions do conscious recognition and projection intervene. Means are central: "It is simply impossible to have an end in view or to anticipate the consequences of some proposed line of action save on the basis of some, however slight, consideration of the means by which it can be brought into existence."[11] Thus the meaning or value of an object changes as a subject reasons about its practical utility. Means and ends thus become a continuum. Every effect becomes a subsequent cause in a stream of events with no final state.

To advocate incremental progress toward achievable ends is a form of pragmatism. The pragmatist desires ends whose properties will serve as further means. This makes the end an object. The achievement of end objects makes design into a problem-solving activity. The utilitarian value of solving a problem by design depends on the recognition of needs in terms of something that is lacking. This works well for consumerism, whose central tenet is that the acquisition of an object can solve or relieve some need or desire. Here then is today's normative construct of adding value: delivering solutions.

Thus in its everyday usage, the word *utility* refers to the objective capacity of something to satisfy wants. This has been one basis for the twentieth century's economics of markets and prices.[12] Within the larger history of value, however, utility more accurately represents satisfaction obtained than the fitness of an object for a purpose. This distinction justifies the current shift from performance to appropriateness in technology design.

This more accurate conception of utility is contextual. As desirability, it depends on momentary needs, competing options, previous consumption, and the like. What some thing or some state of mind can be used for has more value if it agrees with what one desires and knows how to do. Utility seems to be a more general version of "usability," which is the dominant criterion of so much recent computational interaction design.

Even so considered, utility remains an incomplete substitute for value. According to the often-cited warning by the economist Joseph Schumpeter, half a century ago, "the problem of value must always hold the pivotal position, as the chief tool of analysis in any pure theory that works with a rational schema."[13]

When taken to its logical extreme, as it was over the course of the twentieth century, a utilitarian theory of value that favors increments, instrumentality, and means can lose sight of previously acknowledged forms of value in totalities, intrinsic being, and higher ends.

For present concerns in design practice, a solely utilitarian conception of value relegates design to an after-the-fact role in the embellishment of practical solutions toward ends in view. In other words, utilitarian design is seldom given the chance to bring new ends into view, or to influence the means of solutions.

Economism and Placelessness

Beyond its analogy to the current philosophical debate in technology design, this recitation concerns the need for a next evolutionary stage in economic history, of which better-situated interaction design could be a healthy indicator.

The value of place, an otherwise hazy sentiment, provides a good way to approach this. As a discipline, design for the environment seems aware of conditions that fewer economics textbooks explain: our need to escape from the reduction of so many forms of value to market price, and to improve our accounting of nonfiscal value.

A science that accepts a claim that whatever sells must be good is a dismal one.[14] The discipline of economics excuses itself as solely an inquiry into pricing and market mechanisms; but much as the Middle Ages were ruled by theology, modernity has been ruled by this dismal science. Much as medieval discourse had to remain accountable theologically, today every argument must justify itself in terms of quantifiable economic value.

One word for this state is *economism*. Trade may always have been a prerequisite of cultures, but it was seldom as independent of them as it has now become. In a traditional society, any material exchange occurred within stable patterns based on social customs and the rule of church or state. Histories of even the most mercantile peoples were seldom articulated in terms of market forces. Only in a modern society do markets appear at the center of things, with a life of their own, and in tension with traditional institutions. Only then does economics arise as a dominant discipline.[15]

Economism has become an end in itself. Modernity's fiscal regime perhaps began with industrialism's emphasis on material means, but it has become more accurately characterized, especially since the rise of computing and globalization, by its emphasis on abstract monetary ends.

In an increasingly transcultural economy, the one measure upon which anyone can agree, despite other barriers, is money.[16] Besides reducing value to price, economism gives undue emphasis to things measurable with greater precision.[17] Subjectivity inconveniences analysts. To ignore what is not easily measurable lends some irony, however, to the way analysts so proudly declare themselves to be "value free." Any gathering of data involves commitment to a particular set of beliefs.[18]

By the close of the twentieth century, the increase of fiscal capital appeared as a principal goal of industrial societies, no longer merely its means. And what an increase! The rule of economics has yielded a degree of abundance that would previously have been inconceivable even to kings. Because human endeavors have always strived for increase, and because increase remains one of the best explanations of value,[19] modern material abundance has to be understood as a triumph.

Now the costs of economism are becoming more obvious, however. The more that accumulation and convenience become treated as the only basis of human well being, the more people sense that the quality of life has suffered. Any fool can see that in the western world we are drowning in goods, but lacking in good.[20]

This is essentially an accounting problem; not all that humanity has traditionally valued about the world is equally quantifiable. Those phenomena that afford unambiguous measure are more likely to be reflected in the bottom line; and those that are unpredictable or overly complex are more likely to be left out. Moreover, the use of complex computational models gives power to mathematically indicated solutions at a cost to the semantic significance of the operands. Economic increase has been separated from human activity as a result.[21]

Thus, whereas the accumulation of separate things—even redundant, unnecessary, or pernicious things—is always counted as value added, the degradation of interconnected cultural and natural environments remains uncounted as value lost. Representing the value added by the manufacture and sale of shoes is easy, even in a nation where most people have more than enough shoes already, but representing the value of pleasant urban spaces for walking in those shoes is more difficult.

Nonmonetary forms of accumulation are being depleted at an alarming rate. Complexities of culture, nature, and humankind have been overlooked, or even eliminated, in scientific economics' quest for predictability.

In other words, one traditional basis of value undergoing decrease is place. Although this is not the moment to join the debate on globalization, we should be able to admit without controversy that global wealth has come at palpable cost to local value. The proliferating realms of the global economy appear built to serve capital before people. The cultural identities that we hope were once the ultimate ends of human endeavor have been reduced to raw material for the increase of abstract capital. Consuming this raw material has turned public spaces into theme parks. The exhaustible raw material now appears to be places themselves.

Expanding the Measures

The postindustrial agenda of the twenty-first century demands more accurate valuation of habitual living patterns. As in assigning value to biodiversity, in the new century we find "something the matter" in the causes by twentieth century neoclassical economics. In its concern for incremental value and market price, that economics took for granted the total, priceless value of life-sustaining planetary systems (*causa materialis*). In its preference for growth over maintenance, quantity over quality, and extensiveness over intensiveness, it neglected much potential for design (*causa formalis*). In the incompleteness of its accounting, modernist economics blinded itself to the increase or decrease of living systems (*causa finalis*).[22]

Advocates of sustainability have suggested the way toward a more accurate, inclusive economics. The essential accounting problem under which we all suffer is that capital has been addressed only as money and gear. This new century's agenda must expand the measures of value to reflect the increase (or at least maintenance) of capital forms neglected by modernism. This is not some anticapitalist nostalgia, but a more accurate and complete capitalism, based on an expanded representation of value and given leverage through appropriate digital technology.

The economist and systems theorist Paul Hawken has articulated this paradigm shift in terms of life cycle costs and benefits.[23] This more complete accounting is not crippling, but instead shifts economic activity away from production for production's sake toward more life-centered constructs of value. The increase of such value is nevertheless understandable and can be instrumentalized as capital. At the heart of a new economy is the proposition that finance and the means of production are not the only forms of capital.

Human capital consists of abilities and aspirations. Where industrial technology ignored or depleted this form of capital, postindustrial technology can cultivate it. In other words, using computers and networks with a more human outlook can reunite skill and intellect and create a sense of living well through working well.

Cultural capital consists of the many patterns, identities, and social infrastructures that we have explored in detail. Traditionally it has been maintained by government, religion, and social customs. In recent centuries, economics has assumed power over those arrangements. At present, information technology is shaping cultural capital as well. According to the sociologist Manuel Castells, for example, the greatest benefit of the Internet is not in commerce but as a medium for "social codes."[24]

Finally, natural capital consists not only of the life support services provided by the planet but also the value of wonder at something better organized, more diversified, and less wasteful than the best of human artifice. With origins in a world whose carrying capacity had not yet been reached, modernity lacked incentives to treat the atmosphere, oceans, and earth as depletable capital rather than free goods. This is a central point in the sustainability debate.

Design plays an increasing role in this paradigm shift toward representing nonfiscal capital. Advocates of sustainability commonly argue for the importance of design. Developers of pervasive computing would do well to keep this larger agenda in mind. This stage in the development of computing gives new meaning to the phrase long favored by the energy conservationists among advocates of sustainability: "appropriate technology."

The word *appropriate* implies some moral value. Will pervasive computing do anybody any good, in terms of the many forms of value reviewed here, or will it just add to the glut of consumable goods?

The pertinence of all this "green" economics to situated interaction design is this: Appropriate digital technologies should improve the measurement and evaluation of complex, ambient, and dynamic criteria neglected by twentieth century economics. Appropriateness can be addressed through design and is foremost a matter of context. Contexts are best understood as places, which themselves are repositories of human, cultural, and natural capital.

Context as Capital

Like big piles of money, places are accumulations too. Well-made, well-designed, well-lived-in places are repositories for human, cultural, natural, and fiscal capital. Indeed they serve its increase.

The sites where value is deliberated hold value in themselves. Often this is as institutions. For example, a museum is built to exalt art objects, whose reception becomes biased by the fact that they have become museum pieces, which makes attending (or of course curating) a museum into an act of endorsing aesthetic value. Less grandly, a small town's annual flea market is a repository as well. By giving some identity to what sort of objects you will find there, it assumes value in itself; and since it occurs on a regular basis, it assumes value for its existence as much as for its practicality.

Rather than being commodities exchangeable in markets, places are more likely to be the conditions under which markets themselves occur. According to the political economists Jon Logan and Harvey Molotch:

> Places are not simply affected by the institutional maneuverings surrounding them. Places ARE those machinations. . . The real flaw of market-economy schools is that they ignore that markets themselves are the result of cultures. . . Even when compared to other indispensible commodities—food, for example—place is still ideosyncratic. The use of a particular place, for example, provides access to school, friends, workplace, and shops. . . Places have standing, in competition for public resources. Places are vital units of aggregation. Not all in individual consumption. . . Places are communities of fate. Neighborhoods have standing. Location establishes a special collective interests Residents are continuously buying into a neighborhood, for instance. . . "Production services cluster in the largest places, and seem to have the least tendency toward dispersal.[25]

The fixity of places is critical to understanding their value. The fact that contexts cannot be readily converted to other assets distinguishes them from fiscal capital. Places are forms of nonfiscal capital accumulated for the conduct of interactions.

Value as Impetus

Place becomes critical to the success of pervasive computing technologies. This has already become a problem that is costly to ignore. Place also becomes essential to a more inclusive environmental economics. Now those two movements must combine. In the highest view of design's call to action, interactive technology must become instrumental to a natural capitalism. The design economics of place has become a primary agenda for technology. The latest advance in information technology, and by extension in big business, can and must be a pivotal stage in valuing our environments.

The first step in setting this course has to be a readmission of embodiment. Next there must occur a recognition of embodied predispositions. From that it follows that cultural difference and local usage are much larger repositories of value than has been acknowledged to date. From this it follows that we need to find terms by which to measure such value. New forms of pervasive computing (different from the current emphasis on security) should help us implement such measurement. If it is possible to be optimistic about pervasive computing, this may be how.

Because this outlook demands expanded constructs of value, its implementation becomes both good business and a liberal art. For the benefit of anyone still holding twentieth century notions of economic value, this shift can be explained as a matter of diminishing returns. *Homo faber*, the maker of things, usually tends to add value by means of technology. Yet technology cannot become an end in itself. Both philosophy and economics perennially remind us that things have no value in themselves—they only have value for someone. Nothing is of more general value to someone than a roof overhead, something to eat, and a cure for an ailment. Nothing, among all the societal strategies attempted by humankind, has been more useful for meeting basic needs than modern industrialism.

When those material needs have been met, however, continued industrial production and consumption yields only diminishing returns. No amount of material goods to buy can fill the holes left by the decline of other cultural functions such as civics, religion, small-scale farming, or arts and crafts. And so the focus of postindustrial production shifts to entertainments.[26] Like material comforts, this intellectual sustenance is more than welcome.

Yet similarly, when basic needs have been met, continued production of fantasies yields only diminishing returns. No quantity of places to go, games to play, and spectacles to watch can fill the spiritual void. Under these regrettable conditions, the main purpose of technology appears to have become distraction engineering, and the main ambition for the technological future is to turn up the resolution.[27]

Ultimately, this concerns what makes humans human. The ultimate source of value is whatever people delight in doing. It does not have to consist of fulfilling the same need over and over again to ever more highly induced levels. It may include fulfilling more abstract needs that may not have been considered while more basic needs have been getting met. It may involve discovering value in things that were either unattainable or taken for granted by previous human generations. In the end, the design of technology cannot leave us as spectators and consumers, but must let us actively practice at something, however humble. Taking part in a locale is one such activity.

Despite all we love about mobility and flow, the kinds of organizational, cultural, and natural capital now so dear to postindustrial economists are still best found in fixed settings. Quite simply, places remain great accumulators of value. New technologies may undermine such value, or they may recognize and enhance it. Built environments may be enhanced or trivialized, but they are seldom completely replaced by the flux of people, gear, and data. Response to place, now mandatory in the largest possible sense, demands major choices in the contextual design of technology. Can design help us realize the many kinds of capital of which places are repositories? The main practical truth here bears repetition: In a genuinely new economy, what constitutes value itself must change.

IV Epilogue

10　Going Native

You might wish for a simpler life. Filling the world with sensors, effectors, and microchips probably sounds like just the opposite of this. Sites that respond to you may also unnerve you. Objects that interact may never seem as natural as objects that you just operate, like a kitchen sink.

At the usual level of understanding, all technology is unnatural. One of the most common definitions of natural is the absence of artifice. But at another level, understood as our own species' means of adaptation, technology is natural too. In this sense, our homes and highways and management information systems are as natural as the nest a bird builds. They are forms of adaptation within a living ecosystem. According to naturalists who take this position, such as the design advocate Janine Benyus, "the question is not whether our technology is natural, but whether it is well adapted."[1]

Almost all design philosophies are based on a belief in an unseen order, learning the ways of which is the main opportunity for human good. Such ways may be manifest in scientific law, moral code, divine vision, economic efficiency, genetic impetus, or ecological adaptation. In each of these cases, an unseen order exhibits a subtle, inexplicit essence that we refer to as its "nature."

Humanity naturally adapts to being in the world by using technology. The sustainability of our species depends on the appropriateness of our adaptation. Technologies of world making become dangerous unless they are complemented by technologies of world knowing.

The poet Gary Snyder, whose prose essays in *The Practice of the Wild* meditate on place as well as any, suggests adaptation at a continental scale. Identifying with a bioregion rather than a political jurisdiction requires at least some knowledge of the biome, however, and right now, there are too many people who "don't even know that they don't know the plants."[2]

There are tens of millions of people in North America who were physically born here but who are not living here intellectually, imaginatively, or morally. Native Americans to be sure have a prior claim to the term "native." But as they love this land they will welcome the conversion of the millions of immigrant psyches into fellow "Native Americans."[3]

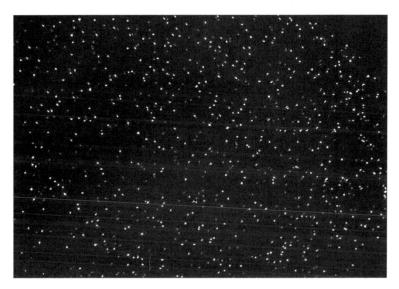

10.1 A stargazing metaphor: faced with immensity (of technologies and communications) feel the value of ground. (*Source: Mt. Wilson observatory star map generator.*)

The expression *going native* apparently originated in nineteenth-century India. It appears in the writings of Kipling, but not in any literature from much earlier. Apparently it began as a description of Englishmen wearing loose-fitting pajamas in public. This sensible adaptation to the sultry climate was seen as a token of deeper assimilations, particularly intermarriage, which the expression came to represent. Such practices were common enough amid mercantile colonization in the eighteenth century, but as foreign traders became rulers, the accompanying social tensions made assimilation taboo. Thus to the imperial British of the nineteenth century, "going native" was a crime. It represented a lapse of discipline and a descent into chaos.

A century later, at least some English-speaking people see local complexities quite differently. Whether as mathematicians, biologists, or urbanists, we recognize how there can be more, not less, rigor in complex, only superficially disordered systems. In a word, we have begun to understand ecologies, and to design within them. As the dynamics of complex natural systems increasingly inform the design of artificial systems, ecological principles have become relevant to human organizations, services, even industries. "Industrial ecology" is no longer an oxymoron.

Rejecting technology outright is not an option. As the hippies learned, going native by retreating into subsistence agriculture generally fails. There is no point in opposing technology per se, as if, say, air conditioners, aspirin, and address servers were all part of one big wrong turn in human history. Even the most left-bank antitechnologist regrets it when the electric power goes out.

Instead, the mimicry of biological systems depends in increases on applied intelligence. Through closer observation and more precise engineering, human artifice copies the resilience and wastelessness of nature. The addition of distributed sensing, local memory, abstract pattern recognition, and feedback control systems advances this imitation.

We have seen how this requires more inclusive economics, and how those are based on realizing local value through design. In this sense, going native can no longer be dismissed as romanticism. In our age of technological saturation, response to place becomes the most practical adaptation strategy of all.

Notes

Introduction

1. This introduction stands on the shoulders of an earlier and more prominent one, which is worth including here. John Thackara introduced the 2002 Doors of Perception conference on pervasive computing as follows:

   > Trillions of smart tags, sensors, smart materials, connected appliances, wearable computing, and (soon) implants, are being unleashed upon the world. To what question are they an answer? What social consequences will follow when every object around us becomes smart and connected?
   >
   > You might think that these questions would preoccupy anyone affected by technology—namely, all of us. But you'd be wrong. Technology was so over-sold during the dot-com boom that now, when boom has turned to bust, we don't even want to think about it.
   >
   > This is a bad mistake, for two reasons. First, because a wave of new technology is coming anyway—and if we don't set the agenda for its use, others will. Police and security forces, and the military, have malign plans for all this stuff. Meanwhile, non-military companies push the concept of "proactive" computing that is "perceptive of our needs"—needs that they reduce, on our behalf, to in-home communication, and entertainment.
   >
   > With a combination of arrogance and ignorance, the designers and companies who promote pervasive computing behave as if the natural world and its inhabitants—you and me—simply did not exist. As the inaugural issue of Pervasive Computing put it this summer, our world is "like the American West...a rich, open space where the rules have yet to be written and the borders to be drawn."
   >
   > Pervasive technologies promise to transform the ways we experience and live in the world. But did anyone ask for our permission? Have we debated the consequences? We have not. After all, tech is old news. (www.doorsofperception.com)

2. These words about the unrealized potential for interaction design echo various position statements from the Interaction Design Institute, Ivrea, Italy, including remarks from the director, Gillian Crampton-Smith.

Chapter 1

1   Brenda Laurel, *Computers as Theater.* Reading: Addison Wesley, 1991. There may not be a more insightful or prescient book about wresting digital media from the engineers.

2   On smart tollbooths: *Scientific American*'s Working Knowledge column in the December 2001 issue, observed that more than seven million Americans have electronic toll tags on their vehicles. The technology works at up to 100 mph; one slows down for safety reasons, not to be scanned. Besides increasing the throughput of the booths, and thereby reducing emissions and delays, this system facilitates microtransactional discounts for offpeak use, etc. Manufacturers are testing tags that use debit card inserts so that accounts are unnecessary, thus protecting privacy. "Stop here once (to sign up), then never stop here again!" advertised one Boston tollbooth using this system.

3   The spread of text was among Mark Weiser's favorite narratives of ubiquity.

4   The Intel statistics are from *Communications of the ACM,* 43(March 2000):3.

5   This is an instance of mainstream technofuturism from an editorial introduction in the digital research commmunity's flagship journal: "It's clear we've reached a turning point in the way we interact with computers... In this [issue] you'll learn of clothes that recognize acquaintances, computers controlled by facial or hand gestures, chairs that adjust to individual bodies, machines that sense the users' moods, and rooms that metamorphose to users' needs." (*Communications of the ACM,* 43(March 2000):3.)

6   Much experience can be understood as interaction, of course. As a usual point of departure, one can approach interaction as a conversation. One party acts or speaks, and the other interprets and then responds. This distinguishes the exchange from mere transmittal. The more that the response follows one assertion and invites another, the more engaging the interaction is thought to be. Supporting contexts and protocols help keep this on track.

7   National Institute for Standards and Technology, "*Pervasive Computing* is a term for the strongly emerging trend," www.nist.gov/pc2001.

8   Intel Corporation, "The Computing Continuum Conference." March 2000, http://www.intel.com/intel/cccon/.

9   MIT Project Oxygen, oxygen.lcs.mit.edu. This was perhaps first announced by Michael L. Dertouzos in, The Future of Computing, *Scientific American,* July 1999.

10   "21 Ideas for the 21st Century," *BusinessWeek,* August 30, 1999.

11 "Through the Looking Glass," the 1989 essay by John Walker, remains the best expression of the belief that the symbolic illusions so successful in the transition from the one-dimensional command line to the two-dimensional desktop-windows interface would next naturally be extrapolated to three dimensions and beyond. Walker was the founder and CEO of Autodesk, the one major technology company that bet most heavily on cyberspace. In Brenda Laurel, ed., *The Art of Human-Computer Interface Design*. Reading: Addison Wesley, 1989.

12 *New York Times,* advertising campaign, December 1996. "Created by a special staff of New York Times editors, reporters, and writers, CyberTimes goes far beyond being just news. It is a compendium of news, columns, on-line forums, in-depth features and guideposts for anyone trying to fully grasp this constantly shifting medium. CyberTimes examines the impact of cyberspace from all perspectives, then presents it in workable, fascinating pieces."

13 "Multimedia Review," *Economist,* June 21, 1997, p. 12.

14 The arrival of practical immersive displays led of lot of people to believe that one no longer had to be a mystic to experience a vision. As is all too familiar a story from the early 1990s, stereoscopic, gesture-driven interfaces—the gloves-and-goggles thing—took the popular imagination by storm, and made "cyberspace" and "virtual reality" into household words. Of this, the best technical manifesto was Steve Aukstakalnis' 1992 book, *Silicon Mirage—The Art and Science of Virtual Reality.* (Berkeley: Peachpit Press), and perhaps the most coherent of the many scholarly treatments was *Cyberspace, First Steps,* which was edited by Michael Benedikt (Cambridge: MIT Press, 1992). Yet to most of us who experienced VR at any length, suspension of disbelief didn't make it to the inner ear. Meanwhile the apparatus was neither dignified, nor sociable, nor comfortable. VR literally got in our face.

15 This is similar to dungeons and dragons, that canonical genre of digital entertainment. Today in the age of satellite imaging and global positioning systems we might have trouble understanding true wilderness. The edges of medieval maps were often inscribed with "*hic sunt leones*" (lions here), and that expression still dredges up some very primitive stages of mind. (Yi-Fu Tuan, *Landscapes of Fear.* Minneapolis: University of Minnesota Press, 1979)

16 In cyberspace, "The first ones in were "desperados and mountain men," as John Perry Barlow put it. Although entrepreneurs soon enough outnumbered the renegades, still a lot of what made cyberspace a world of its own, rather than a part of ours, was *distrust*. William Gibson, to

whom we must attribute the word itself, dressed cyberspace in black. It is all too familiar how cyber was punk. "Case was twenty-four. At twenty-two, he'd been a cowboy, a rustler, one of the best in the Sprawl...jacked into a custom cyberspace deck that projected his disembodied consciousness into the consensual hallucination that was the matrix. A thief, he'd worked for other, wealthier thieves, employers who provided the exotic software required to penetrate the bright walls of corporate systems, opening windows into rich fields of data." (William Gibson, *Neuromancer,* New York: Ace Books, 1984, p. 4.)

17   Even early twentieth-century Paris, an urban delight, had its hit list of *îlots insalubres*—those few remaining islands of disease-breeding conditions which today we only know from *Les Miserables.* Concerning fear of urban darkness, see Norma Evenson, *Paris—A Century of Change, 1878-1978.* (New Haven: Yale University Press, 1979). A few of the *îlots,* such as the quartier Mouffetard behind the Pantheon in the fifth arrondissement, are now tourist destinations.

18   Gibson, *Neuromancer,* p. 51. It is noteworthy that Gibson sued the design software company Autodesk to prevent them from trademarking the word "cyberspace." In this case, the word referred to three-dimensional interfaces. In hindsight it seems odd that up against the very much higher dimensionality, embedding, and at best partial ordering of hyperspace, three dimensions of Cartesian interface were believed to be any better than two.

19   For example, one of computer science's leading cultural consciences, David Gelertner, has devoted a book to his own memories of the fair, 1939, *The Lost World of the Fair.* (New York: Free Press, 1995.) Meanwhile, on television, the fair's imagery which was once popularized as the Jetsons recently appeared as a Simpsons spinoff named for the fair's finest offering, Futurama. A 1980 exhibit and catalog by the Queens Museum in New York opened the gates for deconstructing the fair. Here, according to scholars of the day, was first, best, evidence of architecture's complicity with consumerism.

20   Official Guide to the 1939 World's Fair, p. 29.

21   Lewis Mumford, cited by Joseph P. Cusker in "The World of Tomorrow: Science, Culture, and Community at the New York World's Fair", the lead essay in Helen Harrison, ed., *Dawn of a New Day—The New York World's Fair, 1939/40.* Catalogue of the Queens Museum exhibit. New York: New York University Press, 1980.

22   Sigfried Giedion, *Space, Time, and Architecture—The Growth of a New Tradition.* (Cambridge: Harvard University Press, 1941, p. 432.) The darkness and classical weight of the industrial city was to be overcome

with light and motion. Not only people and their vehicles, but conceptual space itself, now the fundamental concern of the architect, was to flow freely and openly. This ambition has been referred to as modern architecture's project of "transparency."

23   "The road to Tomorrow leads through the chimney pots of Queens," E. B. White, "The World of Tomorrow." This article appeared in The New Yorker in 1939; it is now in print in the collection, *Essays of E. B. White.* (New York: Harper and Row, 1977, pp. 139-148. Quote on p. 139.)

24   John Thackara, "The Design Challenge of Pervasive Computing." Introduction to the conference, Doors of Perception 7: Flow. www.doorsofperception.com, 2002.

25   Thackara, "The Design Challenge of Pervasive Computing."

26   For one among a great many recent books on the privacy question: Simson Garfinkel, *Database Nation: the Death of Privacy in the 21st Century* (Cambridge: O'Reilly, 2000) was a widely read book on how each of us casts a "data shadow" that increases the risk of violations such as identity theft.

27   David Shenk (*Data Smog: Surviving the Information Glut,* San Francisco: Harper Edge, 1997) advised each of us to be our own filter. His argument was less about how well organizations can sort through huge databases than how information pollution causes stress in individual lives.

28   For examples of protest about the current state of interface design, the overemphasis of features and specifications, and the relegation of design to an after-the-fact endeavor, see Alan Cooper, *The Inmates Are Running the Asylum, or, Why High Technology Products Drive Us Crazy and How to Restore the Sanity* (Indianapolis: SAMS, 1999); Terry Winograd, *Bringing Design to Software* (Reading: Addison Wesley, 1996); Jenny Preece, Yvonne Rogers, Helen Sharp, *Interaction Design—Beyond Human-Computer Interaction* (New York: John Wiley, 2002); and other such standards in the field.

29   Henry Dreyfuss, Designing for People. (New York, Simon and Schuster, 1955.)

30   On industrial interfaces: Contact with machinery that processed materials, not symbols, tended to be a brute-force labor condition. Technology was something to turn on, feed material into, watch passively but constantly, and to which to surrender any variations in the pace of one's own life. For these, the key measure of usability was whether a process needed to be attended. Interfaces did away with participation, for which they substituted command and control. The interfaces for these machines were

the province of technicians. As long as computers were inconvenient to use, corporations and professions could carry on the tradition that gentlemen do not operate machinery. Research focused on process efficiency. Usability was a matter of Taylorist time-and-motion economies—such as counting clicks.

31 Deliberation is two-way, but computers are not yet capable of open conversation. Joseph Weizenbaum's pioneering program Eliza demonstrated the popular tendency to assume that interactivity required thought and therefore that interactive computers constitute artificial intelligence. (Weizenbaum, *Computer Power and Human Reason*. San Francisco: Freeman, 1976.) By contrast, Weizenbaum's criterion of artificial intelligence was that a truly intelligent computer device might answer a query with the remark "say, that reminds me of a story…"

32 Clement Mok, *Designing Business*. San Jose: Adobe Press, 1996, p. 128.

33 Paul Dourish has voiced this idea especially well in his philosophical groundwork for interaction design, *Where the Action Is* (Cambridge: MIT Press, 2001). "The question is not whether this or that technological facility will be available to us; the question is how will we be able to understand it, control it, interact with it, and incorporate it into our lives." (p. 193)

34 On how every limit is to be overcome through technology, in a futures forum, the dean of computer science at a leading university could speculate openly about developing technology to offload the human mind onto a newly cloned replacement body—some sort of mind-brain dualist immortality—and not be laughed off the stage. Raj Reddy, "Earthwares" forum, Carnegie Mellon, Pittsburgh, 2000. Except for the late Herb Simon, hardly any speaker mentioned the Earth, for which the conference was named, especially in terms of learning more about the limits it presents.

35 There may be no better story of the imposed abstraction than John McPhee's essay about the US Army Corps of Engineers's attempts to straighten the Mississippi river. McPhee, *The Control of Nature* (New York: Farrar, Straus, Giroux, 1989.)

36 Charlene Spretnak has articulated in detail the position of "Don't call it romanticism!" (*The Resurgence of the Real—Body, Nature and Place in a Hypermodern World*. New York: Routledge, 1997.) My epilogue ends on this phrase, which is Spretnak's.

37 Ian McHarg, Design with Nature. New York: Garden City, 1969.

Chapter 2

1. In his book *Where the Action Is* (Cambridge: MIT Press, 2001), Paul Dourish has explored the topics of this chapter at great length and to considerably greater depth. I have cited his accomplishment and borrowed some of his words, but my work did not really follow from his. I worked in parallel and this chapter was written before his book appeared. The assertion that "embodiment is a property of interactions" is his.

2. Edward Hall, *The Hidden Dimension.* New York: Doubleday, 1966, p. 58.

3. Yi-Fu Tuan, *Space and Place, The Perspective of Experience.* Minneapolis: University of Minnesota Press, 1976, p. 36.

4. From a generation before the current fashion for body-based theory and criticism in architecture, one of the more lasting books was Kent Bloomer and Charles Moore's *Body Memory and Architecture* (New Haven: Yale University Press, 1977.)

5. The Heideggerian principle pervades these ideas. We understand the world mostly in terms of how we can engage it.

6. Byron Reeves and Clifford Naas, *The Media Equation—How People Treat Computers, Television, and New Media Like Real People and Places.* Cambridge, UK: Cambridge University Press, 1996.

7. It is important to recognize that body image remains provisional. Neither complete nor certain, it operates in psychological tension with environments and persons, and with a disconnect between visual symbols and physical memory. An enormous beauty industry induces and then preys on such insecurities, for instance.

8. The human capacity for environmental mental models appears to predate and transcend linguistic description. Spatial representations apparently underlie the intuitive physics and shifting frames of reference by which we interpret everyday life. Engaged actions produce direct perceptions, however, which when shared create a strong sense of environmental possibilities without need for a detailed operational mental model.

9. The movement got under way when people started spending too much time sitting at computers. The number of body art shops keeps climbing. Without some sort of business organization to collect the data, aggregate statistics on the number of people with piercings are rather scarce. The website www.piercinglinks.com provides a good window on the subculture, however.

10. The body is the bearer of cultural symbolism. Where external agencies such as church and state once provided security, now the body itself comes to the foreground as the site of identity. "Writing the body" has

become a popular trope within the discourses of critical theory. See, for example, Chris Shilling, *The Body and Social Theory* (London: Sage, 1993). This approach breaks from previous tendencies to regard body art as an attribute of primitive cultures, or of misguided moderns sampling primitive cultural traits out of context.

11. Recent body art reflects a different attitude than the customs by which more rooted peoples painted themselves. While it may admire and appropriate traditions, it uses them to plumb the contemporary psyche. There it finds a tension between the more traditional distress of embodiment (i.e., the experience of mortality and pain), and the newer distress of disembodiment in a technological world less well mapped to human scale. As Hal Foster pointed out in *The Return of the Real* (Cambridge: MIT Press, 1996), "ob-scene" art works because, lacking a scene it comes too close to us. Or rather, the scene is the body.

12. The pathology of Lacanian psychological schisms? Whatever the wonder in revisiting infantile delight or pubescent embarrassment at body functions, and despite the stereotypical tendency to see repressed trauma where there is none, body art also strikes some more sober chords of distress.

13. Robert Hughes, on the viscerality of Robert Mapplethorpe, "Art, Morals, and Politics," *New York Review of Books*, April 23, 1992.

14. Foster, *The Return of the Real*.

15. Samuel Johnson, from the *Oxford Dictionary of Quotations*. Cited (but buried) in Benjamin Wooley, *Virtual Worlds*, London: Penguin, 1993, p. 244.

16. For a normative account of Descartes' influence on the mind-body problem, see Gregory, Richard, and O.{t}L. Zangwill, eds, *The Oxford Companion to the Mind* (New York: Oxford University Press, 1998, pp. 487–491).

17. Many of us spend the better part of the day staring at a computer screen. As explained to me on a visit to Microsoft Research, the position we assume is jokingly referred to there as "sitting at the altar."

18. Being wet is an emergent property of the structure of piles of H_2O molecules; similarly, according to the "attribute-theory" camp in mind science, thoughts come from the structure of brain process. John Searle, *The Rediscovery of the Mind* (Cambridge: MIT Press, 1992.)

19. George Lakoff and Mark Johnson, *Philosophy in the Flesh—The Embodied Mind and its Challenge to Western Thought*. New York: Basic Books, 1999.

20. "Spatial relations concepts are the at the heart of our conceptual system." (Lakoff and Johnson, *Philosophy in the Flesh*, pp. 30–36.)

21. "[The environment] is not a collection of things that we encounter. Rather, it is part of our being. It is the locus of our existence and identity. We cannot and do not exist apart from it. It is through emphatic projection that we come to know our environment, understand how we are a part of it, and how it is part of us. This is the bodily mechanism by which we can participate in nature, not just as hikers or climbers or swimmers, but as part of nature itself, part of a larger, all-encompassing whole." (Lakoff and Johnson, *Philosophy in the Flesh*, p. 566.)

22. Naomi Eilan, Rosaleen McCarthy, and William Brewer eds, *Spatial Represen-tation—Problems in Philosophy and Psychology*. Oxford: Blackwell, 1993.

23. Eilan, McCarthy, and Brewer, eds, *Spatial Representation*. Categories of issues in spatial thinking are: frames of reference, intuitive physics, spatial representation in the sensory modalities, action, what and where—perception. Categories of questions about spatial representation are the following. (1) How does spatial thought concern the physical environment? (2) How does spatial thought provide a disengaged picture of a persistent world? (3) What about the world we immediately inhabit? (4) How do we view ourselves in the world as if from outside?

24. On mental models: Does any action in an environment require such a representation, or can direct perception sometimes preempt mental models? And what is the role of peripheral, as opposed to deliberative awareness of context? The mind successively interprets raw stimuli to create and refine a mental representation. Because cognitive limitations prevent awareness of all perceptible conditions, stimuli must be filtered. For example you might cease to hear the traffic outside a new residence after a month there. This requires categories, against which any salient, contrasting, or anomalous stimuli can be quickly evaluated for whether they demand a shift of attention.

25. Tacit knowledge was a main focus of my previous book *Abstracting Craft* (Cambridge, MIT Press, 1996). On this subject the best authority is the chemist-philosopher Michael Polanyi, whose book *Personal Knowledge* (Chicago: University of Chicago Press, 1958) remains the best statement of how even objective data gathering is not without personal involvement.

26. When even the psychologists seem afraid to appear introspective, something was wrong. Barbara Tversky, an eminent scholar in psychology, once assured me that Carl Jung, for example, was not a psychologist.

27. For a thorough compendium, see Reginald Golledge and Robert Stimson, *Spatial Behavior—A Geographic Perspective* (New York: Guilford Press, 1997).

28. A. W. Siegel and S. H. White, "The Development of Spatial Representations of Large-Scale Environments," in *Advances in Child Development and Behavior*. New York: Academic Press, 1975. Volume 10, pp. 10–55.

29. Barbara Tversky, "Spatial Mental Models." In *The Psychology of Learning and Motivation*. New York: Academic Press, 1991. Volume 27, pp. 109–145.

30. Kevin Lynch, *The Image of the City*. Cambridge: MIT Press, 1960.

31. The idea of a "cognitive map" dates at least as far back as Tolman's pioneering experiments in the late 1940s, in which rats climbed out of mazes they had mastered in order to take shortcuts to food. Perhaps the best expression of the academic fashion for cognitive maps was assembled by Roger Downs & David Stea in *Image and Environment* (Chicago: Aldine Press, 1973). More recent scholars agree that while spatial mental models are like maps, or are at least used like maps, nevertheless they are not exact maps.

32. Phenomenology has been one of the more significant counteractions to have emerged from the high modernity that resulted from mind-body and culture-nature dualisms. Phenomenology responded to mechanized abstraction with a renewed focus on presence. Heidegger emphasized dwelling. Artists such as Brancusi appealed to archetypes of organic form. More recent "biological naturalism" challenges many of the working assumptions of more mechanistic theories of the mind-body split.

33. Maurice Merleau-Ponty, *Phenomenology of Perception,* Colin Smith, trans., London Routledge, 1962, p. 203.

34. Merleau-Ponty, *Phenomenology of Perception,* p. 440. Cited and interpreted by Hubert L. Dreyfus in "The Current Relevance of Merleau-Ponty's Phenomenology of Embodiment". Dreyfus, Hubert. "The Current Relevance of Merleau-Ponty's Phenomenology of Embodiment," in Honi Haber and Gail Weiss, eds, *Perspectives on Embodiment,* New York: Routledge, 1996, pp. 103–120.

35. In a recent defense of Merleau-Ponty, the philosopher Hubert Dreyfus, a perennial critic of mechanistic, information-processing models of humankind, has stated: "These three ways our bodies determine what shows up in our world—innate structures, basic general skills, and cultural skills—can be contrasted by considering how each contributes to the fact that to Western human beings a chair affords sitting. Because we have the sort of bodies that get tired and that bend backwards at the knees, chairs can show up to us—but not flamingos, say—as affording sitting. But chairs can only solicit sitting once we have learned to sit. Finally, only because we Western Europeans are brought up in a culture

where one sits on chairs do they solicit us to sit on them. Chairs would not solicit sitting in traditional Japan. By embodiment, Merleau-Ponty intends to include all three ways the body opens up a world." (Dreyfus, "The Current Relevance of Merleau-Ponty's Phenomenology of Embodiment")

36. For a normative account of Husserl's influence on the mind-body problem, see Gregory and Zangwill, eds., *The Oxford Companion to the Mind*, pp. 615, 326.

37. In a philosophical summary in the chapter "Being in the World: Embodied Interaction," Paul Dourish has traced its roots to Husserl. "It began with the phenomenologists. I outline Husserl's attempts to reorient the Cartesian program around the phenomena of experience; Heidegger's reconstruction of phenomenology around the primacy of being-in-the-world; Schutz's expansion of the phenomenological program to account for problems of social interaction; and Merleau-Ponty's elaboration of the role of the body in perception and understanding." (Dourish, *Where the Action Is*, pp. 124–125.)

38. J. J. Gibson, *The Ecological Approach to Visual Perception*. Boston: Houghton Mifflin, 1979. This book ranks among a handful of the foundational works for the present state of interactive media.

39. Learning embodiment is what so much of child's play is about. This is one reason why children need so much protection; for example, at some stage they literally do not know that they must not walk off the edge of a cliff.

40. The geographers Gary Moore and Reginald Golledge provided one of the better balanced, if less recent overviews of spatial learning in their edited collection, *Environmental Knowing—Theories, Research, Methods,* (Stroudsburg, PA: Dowden, Hutchison, Ross, 1976). They set out four categories: information processing, personal constructs, environmental learning (i.e., the cognitive), and developmental stages.

41. Not surprisingly, early interest in environmental cognition accompanied the structuralist movement. A major theorists of this movement was the psychologist Jean Piaget, one of whose key assertions was that development is transformative. In support of our more general assertion of abstract spatial relations as a basis for category and metaphor, however, note that the explicit Euclidean measure we normally expect of maps comes only at the last of Piaget's four stages, namely, the formal-operational. The preceding, or concrete-operational stage involves a more topological approach to environmental cognition. Moreover these topologies require the projection of the subject into the environment, that is, the ability to see oneself in place, from which this discussion began.

42. The anthropologist Tim Ingold examined affordances thoroughly in *The Appropriation of Nature—Essays on Human Ecology and Social Relations* (Manchester, UK: Manchester University Press, 1986).

43. Joseph Campbell, *The Power of Myth*. New York: Doubleday, 1988.

44. Civic legibility, of which this is an example, requires some shared spatial literacy. This depends on a shared iconography of good city form. Civic buildings were traditionally the most lavishly constructed buildings, and they occupied the favored, landmark-forming locations within the fabric of the city. For example the post office would be a grand edifice on a public square.

45. Foucault as discussed by Robert Mugerauer, *Interpretations on Behalf of Place* (Albany: State University of New York Press, 1994, p. 17.)

46. Hall, *The Hidden Dimension*, p. 9.

47. Hall, *The Hidden Dimension*, p. 128.

48. Tuan, *Space and Place*, p. 61.

49. In traditional cultures, land ownership was the main basis of wealth and war. In bourgeois cultures, having and marking a place of one's own becomes a prime mover of individuals. Personal territory provides freedom and privacy and it expresses social standing. Unfortunately this form of territoriality leads to the "tragedy of the commons," in which, like herdsmen overgrazing their fields, the pursuit of their best interests by each individual destroys the resources shared by all. (Garrett Hardin, "The Tragedy of the Commons," in *Science*, V. 162(1968), pp. 1243–1248.) The slogan of "Don't tell me what to do with my land!" was the refrain by which James Kunstler in *The Geography of Nowhere* (New York: Touchstone, 1993) criticized the bourgeois version of the crisis of the commons.

50. The Heideggerian fourfold construct consists of: (1) an earth—supporting resource, inscribed upon by our infrastructures; (2) sky—circadian rhythms, manifesting the divine; (3) divinities—manifestation of the hidden, epiphanies in specific phenomena, such as places and events; and (4) mortals: "We are human only insofar as we are mortals, that return to the earth. Bodies are place, where being appears." (Mugerauer, *Interpretations on Behalf of Place*, p. 74.)

51. Christian Norberg-Schulz, *Genius Loci—Towards a Phenomenology of Architecture*, New York: Rizzoli, 1983, p. 19.

52. George Lakoff and Mark Johnson, *Metaphors we Live By*. Chicago: University of Chicago Press, 1980.

53. Mircea Eliade, *The Sacred and the Profane*. New York: Harcourt Brace, 1959. Cited by Mugerauer, in *Interpretations on Behalf of Place*, p. 57.

54. "Moral geography" has been dramatized by the historian Simon Schama, who used it to open his path-breaking history of the Dutch, *The Embarrassment of Riches—An interpretation of Dutch culture in the Golden Age* (New York: Knopf, 1987). There the canonical image was that of the "drowning cell": punishment for indolence consisted of a bail-or-drown setup that reenacted the nation's fundamental struggle to rise above sea level.

55. Cicero, *de Finibus*, vol. 5. Cited in H.{t}J. Rose, *A Handbook of Greek Mythology*, New York: E.{t}P. Dutton, 1959, p. 254.

56. This is an extension to the concluding point from Winifred Gallagher's widely read book, *The Power of Place—How Our Surroundings Shape Our Thoughts, Feelings, and Actions.* (New York: Poseidon Press, 1993). That book demonstrated the extent to which the performance of a building influences well-being. Its concluding point was that people need to become as vigilant about environmental quality on a larger scale as they are about their homes. Given the clear mapping from body to home, as set out by Clare Cooper Marcus, long a visible proponent of the form of participatory design that architects call "post-occupancy evaluation", in her book, *House as Mirror of Self* (Berkeley: Conari Press, 1995) said "I think it fair to map from body to home and on up to region."

57. Even in terms of physical scale, Man is no longer the measure of all things. The dimensions of human endeavor have expanded from body-based cubits and *milia* (a mile was a thousand paces of a Roman legion) to incomprehensibly tiny angstroms and incomprehensibly large light-years. Architecture, a discipline comfortably situated in the middle of this spectrum, and rarely departing from human dimension by more than one or two orders of magnitude, has correspondingly lost authority.

58. "The expansion of our actual identity requires greater recognition of our sense of internal space as well as of the space around our bodies. Certainly if we continue to focus radically on external and novel experience and on the sights and sounds delivered to us from the environment to the exclusion of refining and expanding our primordial haptic experiences, we risk diminishing our access to a wealth of sensual detail developed within ourselves—our feeling of rhythm, of hard and soft edges, of huge and tiny elements, of openings and closures, and a myriad of landmarks and directions which, if taken together, form the core of our human identity." (Bloomer and Moore, *Body, Memory, and Architecture*, p. 44.)

59. "The Ecological Crisis as a Crisis of Character," Chapter 2 in Wendell Berry, *The Unsettling of America*. San Francisco: Sierra Club Books,

1977, pp. 17–26. Quote on p. 22. "The good of the whole of Creation, the world and all its creatures together, is never a consideration because it is never thought of; our culture now simply lacks the means for thinking of it. It is for this reason that none of our basic problems is ever solved." (p. 22) "The possibility of the world's health will have to be defined in the characters of person as clearly and urgently as the possibility of personal 'success' is now defined." (p. 26)

60. Le Corbusier and Francois Pierrefeu. 1942. *Maison des Hommes*. Paris: Plon.

Chapter 3

1. The distinction between outer-directed and inner-directed architecture may have first been made by the cultural landscape historian J. B. Jackson in *Discovering the Vernacular Landscape* (New Haven: Yale University Press, 1984).

2. "The human being by his mere presence, imposes a schema on space." (Yi-Fu Tuan, *Space and Place: The Perspective of Experience*. Minneapolis: University of Minnesota Press, 1976, p. 36.) A generation later, this has become a central theme of much recent architectural theory.

3. This oft-attributed remark paraphrases an argument in the oft-reproduced 1935 essay by Walter Benjamin, "The Work of Art in the Age of Mechanical Reproduction" (in Hannah Arendt, ed, *Illuminations*. New York: Schochen Books, 1969.)

4. For an excellent survey of space itself, with emphasis on the antecedents for recent astrophysics, see Steven Hawking and Roger Penrose, *The Nature of Space and Time* (Princeton: Princeton University Press, 2000.)

5. Henri Lefebvre, *The Production of Space*. Oxford: Blackwell, 1991, p. 15.

6. Manuel Castells, *The Informational City*, Oxford: Blackwell, 1989, p. 169. Castells argued that the changing relation between centralization and decentralization is of greater interest than decentralization itself.

7. Jef Raskin, "Locus of Attention" in *The Humane Interface: New Directions For Designing Interactive Systems*. Reading: Addison Wesley, 2000. pp. 17–31. Quotes on pp. 17, 24.

8. As a result of the foreground being full, mainstream practices of human-computer graphical interface design have reached their limits. As Don Norman has observed, "Making everything visible is great when you have only twenty things. When you have twenty thousand, it only adds to the confusion." (*The Invisible Computer*. Cambridge: MIT Press, 1998, p. 74.)

9. John Seely Brown and Mark Weiser, "The Coming Age of Calm Techno-logy." (www.ubiq.com/hypertext/weiser/acmfuture2endnote.htm, 1996).

10. Brown and Weiser. "The Coming Age of Calm Technology."

11. Bonnie Nardi, "Activity Theory and Human-Computer *Interaction,*" in Nardi, ed., *Context and Consciousness: Activity Theory and Human-Computer Interaction.* Cambridge: MIT Press, 1996, pp. 7–16. Quote on p. 11. This work provides a good overview of contrasting theories of activity and their pertinence to technology design.

12. Nardi, "Activity Theory and Human-Computer *Interaction,*" p. 11. "Activity theory is a powerful and clarifying descriptive tool rather than a strongly predictive theory. The object of activity theory is to understand the unity of consciousness and activity. Activity theory incorporates strong notions of intentionality, history, mediation, collaboration and development in constructing consciousness. Activity theorists argue that consciousness is not a set of discrete disembodied cognitive acts (decision making, classification, remembering...) and certainly it is not the brain; rather consciousness is located in everyday practice: you are what you do. And what you do is firmly and inextricably embedded in the social matrix of which every person is an organic part. This social matrix is composed of people and artifacts. Artifacts may be physical tools or sign systems such as human language. Understanding the interpenetration of the individual, other people and artifacts in everyday activity is the challenge activity theory has set for itself." (p. 7)

13. Lucy Suchman, *Plans and Situated Actions.* Cambridge, UK: Cambridge University Press, 1986. Cited by Bonnie Nardi in "Studying Context: A Comparison of Activity Theory, Situated Action Models, and Distributed Cognition," *Context and Consciousness,* pp. 69–102. Quote on p. 71.

14. Nardi, "Studying Context," p. 71. This work provides a good overview of contrasting theories of activity and their pertinence to technology design.

15. Again, relative to all the usual tossing about of the term affordances, a good explanation exists in Tim Ingold, *The Appropriation of Nature. Essays on Human Ecology and Social Relations* (Manchester, UK: Manchester University Press, 1986).

16. Nardi, "Activity Theory and Human-Computer *Interaction,*" p. 14.

17. Dourish, *Where the Action Is,* p. 158.

18. Dourish, *Where the Action Is,* pp. 100–103.

19. Krippendorf, Klaus, "On the Essential Context of Artifacts, or on the Essential Proposition that Design Is Making Sense Of Things." In Richard Buchanan, ed., *The Idea of Design.* Cambridge: MIT Press, 1995.

20. The ecologist Wendell Berry asserted that America still builds as if it were underpopulated, whereas what it needs most is to learn from much more densely settled places like Japan about the design of courtyards, private gardens, and fences (*The Unsettling of America*, San Francisco: Sierra Club Books, 1977).

21. Christopher Alexander, *The Timeless Way of Building*, New York: Oxford University Press, 1979; and Christopher Alexander, Sara Ishikawa, and Murray Silverstein, with Max Jacobson, Ingrid Fiksdahl-King, and Shlomo Angel, *A Pattern Language—Towns, Buildings, Construction.* New York: Oxford University Press, 1977.

22. Alexander, *The Timeless Way of Building*, New York: Oxford University Press, 1979, p. 65.

23. Alexander, *The Timeless Way of Building*, p. 92. "Of course, the pattern of space does not 'cause' the pattern of events. Neither does the pattern of events 'cause' the pattern in the space. The total pattern, space and events together, is an element of people's culture." p. 94.

24. Alexander, *The Timeless Way of Building*, p. 98. The fact that Alexander's work has been kept in the discourse so long may be due less to architects, whose cult of creativity makes them shun its message, than to computer scientists, whose search for decidable formal systems attracts them to its title. Yet the language fails as a formal method, and the idea that such a system would be desirable at this scale has failed as well. The work seems better taken as a manual on the Tao of traditional architecture. As such, it had relatively little to say about the transformative effects of new technologies. Its lasting relevance instead lies at a somewhat less literal level-in the significance of its many examples toward for we might now call affordances in peripheries.

25. "Fixity" complements "fluidity." Within the context of digital technology, Xerox PARC ethnographer David Levy is usually credited for these terms. ("Fixed or Fluid? Document Stability and New Media." In *European Conference on Hypertext Technology '94 Proceedings.* Edinburgh: ACM, 1994.) Former PARC director John Seely Brown has explained: "There are good cultural reasons to worry about the emphasis on fluidity. But fixity serves other purposes. As we have tried to indicate, it frames information. The way a writer and publisher physically present information, relying on sources outside the information itself, conveys to the reader much more than the information alone. Context not only gives people what to read, it tells them how to read, where to read, what it means, what it's worth, and why it matters." (John Seely Brown and Paul Duguid, *The Social Life of Information*, Cambridge: Harvard Business School Press, 2000, p. 201.)

26. Mary Douglas, "Institutions Confer Identity," in *How Institutions Think.* Syracuse: Syracuse University Press, 1986, pp. 55–67. Quote on p. 67.

27. Douglas, "Institutions Confer Identity." "The whole approach to individual cognition can only benefit from recognizing the individual person's involvement with institution-building from the very start of the cognitive enterprise." (p. 67)

28. Douglas, "Institutions are founded on analogy," in *How Institutions Think*, pp. 45–53. Quote on p. 48. "How a system of knowledge gets off the ground is the same as the problem of how any collective good is created. In Durkheim's view the collective foundation of knowledge is the question that has to be dealt with first. According to his theory, the elementary social bond is only formed when individuals entrench in their minds a model of the social order. He and Ludwik Fleck invited trouble when they spoke of society behaving as if it were a mind writ large. It is more in the spirit of Durkheim to reverse the direction and to think of the individual mind furnished by the society writ small." (p. 45)

29. Even a generation ago, some architects perceived an environmental crisis at the level of embodiment in legible building. "If architecture the making of places is as we propose a matter of extending the inner landscape of human beings into the world in ways that are comprehensible, and if the architectural world is rich in instances of this success, what then is so dramatically wrong with the way we build today? What is missing from our dwellings today are the potential transactions between body, imagination, and environment." (Kent Bloomer and Charles Moore, *Body, Memory, and Architecture.* New Haven: Yale University Press, 1977, p. 105.)

30. In *Repairing the American Metropolis* (Seattle: University of Washington Press, 1997), Douglas Kelbaugh made direct connection between typology and embodiment. "Perhaps a more easily understood example of type and model is the human body. The human being is a single biological species, but it keeps producing miraculous variety.... Not only are subtle differences appreciable, humans do not tire of looking at one another." (p. 111)

31. Kelbaugh, *Repairing the American Metropolis*, p. 109. This book provides a thorough explanation of architectural typology, and especially its role in urban design.

32. Donlyn Lyndon and Charles Moore, *Chambers for a Memory Palace.* Cambridge: MIT Press, 1994.

33. Stewart Brand, *How Buildings Learn—What Happens after they are Built*, New York: Penguin, 1994, p. 54. While nobody would deny the potential of artistry in building, many might agree with Brand that trouble begins with architects' tendency to regard themselves as artists: "The

problems of 'art' as architectural aspiration come down to these: —Art is proudly nonfunctional and impractical. —Art reveres the new and despises the conventional. —Architectural art sells at a distance." (p. 54)

34. Brand, *How Buildings Learn*, p. 54. "Art must be inherently radical, but buildings are inherently conservative. Art must experiment to do its job. Most experiments fail. Art costs extra. How much extra are you willing to pay to live in a failed experiment? Art flouts convention. Convention became conventional because it works."

35. Aldo Rossi, *The Architecture of the City*, Cambridge: MIT Press, 1982, p. 29.

36. Rossi, *The Architecture of the City*, pp. 41, 47.

37. David Nye, *Electrifying America—Social Meanings of a New Technology, 1880-1940.* Cambridge: MIT Press, 1992. There is no more usual analogy to pervasive computing than early twentieth-century electrification, and there is no better study of social and technological change under electrification than this one.

38. John Stilgoe, *Metropolitan Corridor—Railroads and the American Scene.* New Haven: Yale Press, 1983. This excellent volume on the relationship of infrastructure and cultural geography used the railroad as the best case, and Manhattan Transfer and Grand Central Station as its opening studies.

39. In his well-known and still-readable introduction to *Technics and Civilization* (New York: Harcourt, Brace, 1936), Lewis Mumford explained how the diurnal rhythms of monastic life prepared European sensibilities for the strict ordering of time necessary for building a market culture and conceiving an industrial age. Factory labor was sold time, and life in the industrial city demanded rigorous synchronization. Infrastructures of timekeeping occupied a significant place in the architecture of the city; often the bell tower was the most prominent element of a town square, a church, or a factory. (And to this day, amid a much more casual, multicultural, and 24-hour, 7-day way of life, perhaps the one custom still most generally shared in the public presentation of self is the wearing of a watch. Why else would anyone be willing to leave the house in a $15 tee shirt and a $5000 Rolex?)

40. William Mitchell, "Recombinant Architecture," in *City of Bits.* Cambridge: MIT Press, 1996, pp. 47–105. "Type by type, mutations are evident." The types examined include bookstores, libraries, museums, theaters, schools, hospitals, prisons, banks, trading floors, retail stores, workplaces, and the home. Many of these types are dissolved, destabilized, or at the very least overlaid in new ways.

Chapter 4

1. This is everyone's standard citation on the coinage of "ubiquitous computing": Mark Weiser, 1991. "Future Computers." In *Scientific American* 265:3. Special Issue: *The Computer in the 21st Century*, September 1991.

2. For a summary, see Claire Tristam, "The Next Computer Interface," *Technology Review*, December 2001.

3. Don Norman, (1998), *The Invisible Computer,* p. 74. "Although the computer has changed dramatically since the 1980s, the basic way we use it hasn't. The Internet and World Wide Web give much more power, much more information, along with more things to lose track of, more places to get lost in. More ways to confuse and confound. It's time to start over."

4. Don Norman, 1998, *The Invisible Computer*, p. 69.

5. Alan Cooper, *About Face: the Essentials of Interface Design,* Foster City, CA: IDG, 1995.

6. Norman, *The Invisible Computer,* pp. 69, 52. Norman was incidentally also the one to popularize the word *affordances*.

7. Weiser interview in Jennifer Ernst, et al., eds., *The PARC Story*. Palo Alto: Xerox, 1995.

8. Norman, *The Invisible Computer.*

9. Information appliances can network in an ad-hoc manner using protocols that are slimmer and more interoperable than the full-blown Internet TCP/IP.

10. Alex Pentland, "Perceptual Intelligence," *Communications of the ACM,* 43:1, March, 2000.

11. This list pays homage to Steve Shafer's seminal (and much, much more authoritative) "Ten Dimensions of Ubiquitous Computing," which has appeared in various versions including the keynote to Paddy Nixon, Gerard Lacey, and Simon Dobson, eds., *Managing Interactions in Smart Environments (MANSE '99)* London: Springer-Verlag.

12. For a concise survey of web-ready embedded device engineering at new century, see Gaetano Boriello and Roy Want (former colleague of Mark Weiser): "Embedded Computation Meets the World Wide Web." *Communications of the ACM,* 43 (May 2000) : 5.

13. "Smart dust": Berkeley sensor & actuator lab: bsac.eecs.berkeley.edu/research/.

14. DARPA figures given by Intel VP of research, David Tennenhouse, "Proactive Computing." *Communications of the ACM,* 43 (May 2000): 5.

15. The series "Embedded Systems Conferences," launched in 1998, was booming by 2001. www.esconline.com.

16. Paul Saffo, "Smart Sensors Focus on the Future," interview in *CIO Insight*, April 15, 2002.

17. "Affordable smog sensing," *Design Engineering*, 4/25/2002.

18. www.memsindustrygroup.org.

19. "Sensing technology steams ahead." *Electronic News*, 3/18/2002.

20. "Fresh Noise." E-Mag, 5/2/2002.

21. This remark by Scott McNealy at a press conference in January 1999 quickly became a meme on the Internet, where it still circulates.

22. Dallas Semiconductor TINI web server, for instance.

23. Boriello and Want summarized the advantages of the Java model as a refinement over HTML/CGI web-based communications in: "Embedded Computation Meets the World Wide Web." Communications of the ACM 43 (May 2000) : 5.

24. Dan Siewiorek. From a panel discussion "The Computing Continuum Conference," Intel Corporation, March 2000, http://www.intel.com/intel/cccon/.

25. Tennenhouse, "Proactive Computing."

26. Yarin, Paul, and Hiroshii Ishii, (1999), "TouchCounters: Desiging Interactive Electronic Labels for Physical Containers." *Proceedings of Conference on Human Factors in Computing Systems (CHI '99)*.

27. Ishii, Hiroshi and Brygg Ullmer, "Tangible Bits: Towards Seamless Interfaces between People, Bits and Atoms," *Proceedings of Conference on Human Factors in Computing Systems (CHI '97)*, Atlanta: ACM, March 1997, pp. 234–241. This is regarded as the keystone paper on Ishii et al.'s profuse pioneering in embedded gear.

28. www.ibutton.com.

29. CCN Sci-Tech report, March 11, 2002. "Hong-Kongers to get smart ID cards."

30. Sophisticated sensor-actuator systems are just some of the ways in which contemporary culture ironically spends as much as possible on its cars and as little as possible on its buildings.)

31. Michael Fox and Bryant Yeh, "Intelligent Kinetic Systems in Architecture," in Nixon, Lacey, and Dobson, eds., *Managing Interactions in Smart Environments (MANSE '99)*. London: Springer-Verlag, 2000.

32. Enrique Norten, TEN Arquitectos, Educare Gymnasium. *Architectural Record*, 189 (6), June 2001, p. 118.

33. Santiago Calatrava, Quadracci Pavilion, Milwaukee Museum of Art. *Architectural Record*, 190 (3), March 2002, p. 92.

34. Paul Hawken, Amory Lovins and Hunter Lovins. *Natural Capitalism— Creating the Next Industrial Revolution* (Boston: Little, Brown, 1999)

offered numerous shock stories on waste that stems from lack of feedback control systems. In an extreme case, one company which recognized an anomaly in its electricity costs traced excess usage to a heater under its parking lot, which while intended for occasional use to melt snow had been left on all year.

35. Again, a connection to perceptual interaction and skilled participation. The importance of practice was an important theme in my previous book, *Abstracting Craft*.

36. In educating interaction designers to question so much clicking, one curriculum challenges students to design a television remote control that has no buttons at all. First off, this gets students to reconsider, from an ethnographic perspective even, an activity that everyone thinks they know. Second, it raises questions of practicality: what kinds of embedded gear should be feasible the kinds of product volumes suggested by a TV remote. Third, it provokes questions about diversifying interaction: who says all these millions of devices need to be usable in the same least-common-denominator way? Cannot some of them be more enjoyable at the cost of a little more practice? What aspects of ambient, haptic, positional, and collaborative control are worth exploring? What do they say about the activity rather than the object?

37. The EuroHaptics conference. www.reachin.se.

38. For some reason, the decal is generally reserved for the name of a university: apparently you have to be college educated to be dumb enough to put text right in your rear-view line of sight.

39. Jun Rekimoto and Masanori Saitoh, 1999, "Augmented Surfaces: A Spatially Continuous Work Space for Hybrid Computing Environments." *Proceedings of Conference on Human Factors in Computing Systems (CHI '99)*

40. Claudio Pinhanez: "Everywhere Displays." www.research.ibm.com.

41. In *The Media Equation—How People Treat Computers, Television, and New Media Like Real People and Places*. (Cambridge, UK: Cambridge University Press, 1996) Byron Reeves and Clifford Naas went so far as to suggest that scale relationships shape how much "personality" we attribute to artificial displays.

42. Hiroshii Ishii, C. Wisnecki, S. Brave, A. Dahley, M. Gorbet, B. Ullmer, and P. Yarin, "ambientROOM: Integrating Ambient Media with Architectural Space," *Proceedings of Conference on Human Factors in Computing Systems (CHI '98)*, New York: ACM, pp. 173–174.

43. Guimbretière, François, Maureen Stone, and Terry Winograd. 2001. "Fluid Interaction with High-Resolution Wall-Size Displays."

(www.graphics.stanford.edu). At Stanford Interactivity lab, note the architectural intentions. "In creating systems and applications on computers, we are not primarily creating programs, or even creating interfaces to programs. Just as an architect creates a space for living—not just a collection of walls, floors, and windows—the software designer creates a *user experience*. It is in the design of this experience that fluency can be achieved. Design concerns go beyond a particular interface or interaction device, to encompass the setting, the user's background, and the interaction with other people." (interactivity.stanford.edu/ theory.html)

44. Per Enge, "Locator madness pervades plenty of devices." CNN Sci_tech. March 28, 2002.

45. SiRf GPS chips. http://www.abcnews.go.com/sections/tech/DailyNews/ gpsphone980821.html. "Global Positioning Technology Allows Precise People-Finding Cell Phones to the Rescue."

46. http://explorezone.com/archives/00_01/06_ns_tracking.htm.

47. Several of these were pursued in my workshops at Carnegie Mellon, in 1999 and 2000: "Place Identity in Digital Productions."

48. GIS is hardly just software to automate the drawing of maps, as it is commonly misunderstood. It does produce maps as responses to spatial database queries, and it can use maps as query terms, and combine such queries into powerful methods of spatial analysis. Organizations often select their sites, and agencies assess environmental impacts, on the basis of such databases and analyses. Geographic information systems may already lead computer applications (other than spreadsheets, anyway) in their effect on the built environment. Now they become available to millions, and available on site.

49. So far as I know, this expression first appeared around Carnegie Mellon in 1999. UML is already taken for Unified Modeling Language. XML, extended markup language, was all the rage for the web future.

50. Geography Markup Language (GML), opengis.net/gml/00-029/GML.html.

51. Brenda Laurel, *Computers as Theater*. Reading: Addison Wesley, 1991.

52. Shafer, "Ten Dimensions of Ubiquitous Computing."

53. Anind K. Dey, Gregory Abowd, and Daniel Salber. 2001. "A Conceptual Framework and a Toolkit for Supporting the Rapid Prototyping of Context-Aware Applications. Anchor article in Tom Moran and Paul Dourish, eds. *Special issue: Context-Aware Computing, V.16, n 2–4 of Human Computer Interaction*. Mahweh, NJ: Lawrence Erlbaum Associates.

Chapter 5

1. For a snapshot of technical issues in location modeling at the time of this writing, see Michael Biegl, Phil Gray, and Daniel Slaber, eds. (2001), *Proceedings of the Workshop on Location Modeling. Ubicomp 2001.* www.teco.edu/locationws/final.pdf.

2. William Mitchell, *City of Bits—Space, Place, and the Infobahn.* Cambridge: MIT Press, 1996.

3. "Distance is dying; but geography, it seems, is still alive and kicking." ("Putting it in its Place," *Economist,* August 11, 2001.)

4. The case for geometry, and the respective roles of metric and topological representations, has been put best by Steve Shafer and Barry Brumitt of Microsoft Research. Some of the arguments listed here follow from their position, some are common wisdom or common sense, and all of this is at a high enough level to be shaping discourse in the new field of location modeling.

5. Barry Brumitt, John Krumm, Brian Meyers, Steven Shafer. 1999. "EasyLiving: Ubiquitous Computing & The Role of Geometry." Microsoft Research, 1999. research.microsoft.com/easyliving.

6. Brumitt, Krumm, Meyers, Shafer, "EasyLiving: Ubiquitous Computing & The Role of Geometry."

7. Michael Weiss, *The Clustering of America.* New York: Harper and Row, 1988, p. xi. Shortly after the introduction of ZIP codes, the comic strip *Peanuts* introduced a character named 5, whose last name was 94572, and Snoopy wondered, "I never get names straight: did he say 5 or V.?

8. Weiss, *The Clustering of America*, p. xi.

9. Weiss, *The Clustering of America*, p. xii.

10. As evidence of how robust construction site monitoring has become, or at least of how it is used in foul weather, EarthCam's service offering, ConstructionCam, includes a "windshield wiper". www.webcamstore. com/solutions/systems/index.cfm/ConstructionCam.

11. Shoshana Zuboff, *In the Age of the Smart Machine—The Future of Work and Power.* New York: Basic Books, 1988; and Norbert Wiener, *Cybernetics; or, Control and Communication in the Animal and the Machine.* Cambridge: Technology Press, 1948.

12. Search the Internet for "intelligent buildings" and most items found will concern better climate control or broadband network access. Meanwhile "High tech architecture" is mostly just a stylistic idiom involving precision finishes and highly-engineered materials. It is still designed in one-off works, constructed by the same ancient syndicated trades, bought and

sold as square footage, and inhabited with less interest than is given to the desktop computer systems it houses. Security and lighting systems may recognize our presence, but this is hardly the stuff of technofuturism.

13. Lisa Heschong's oft-cited "little book" *Thermal Delight in Architecture* (Cambridge: MIT Press, 1979) argued against the universal uniformity of hermetically sealed, totally controlled environments, which not only wastes energy but also impoverishes life. This argument is still necessary. For example, even in the near-perfect weather of southern California, owners tend to seal off their buildings and set the thermostat at a fixed temperature rather than living with the climate.

14. Francis Duffy, *The New Office*. London: Conran, 1997.

15. "Digital Tools for Age-Smart Housing." *Architectural Record*, Vol. 190, No. 7, July 2002, p. 161.

16. Mitchell, in *City of Bits,* gave plentiful examples of how the timeless Vitruvian triad of "commodity, firmness, and delight" has been applied to the software of places.

17. Sooner or later, many descriptive models become prescriptive. Effective models lead their creators to believe they understand some condition well enough to modify or duplicate it. Furthermore a model may represent a reality that does not yet exist. In this manner, any purposeful activity that creates representations of desired realities constitutes an act of design.

18. Denis Wood, *The Power of Maps* (New York: Guilford Press, 1992), remains a favorite work on the inescapably rhetorical nature of information design.

19. Hilary Putnam, *Representation and Reality,* Cambridge: MIT Press, 1988, pp. 30–32. "Imagine two planets identical except in the molecular structure of the substance they call water. Recall that on earth, 'water' was long thought to be a pure element. That it is H_2O is a fairly recent discovery. On both of these planets any sample of this substance behaves the same as any other sample. Prior to the discovery of chemical compounds, the behaviors on either planet are identical. The mental representations are identical, yet the reality is not. Eventually the compound is known, and experiments are devised in which H_2O on one planet reacts differently than XYZ on the other." (p. 30)

Chapter 6

1. Some of the earliest demonstrations of perceptive computing and "natural" interaction were fixed contexts. The first smart space is generally acknowledged to have been Myron Kreuger's "Glowflow," which was an

installation at the University of Wisconsin in 1969. "Put-That-There," Richard Bolt's landmark demonstration of a wall-sized map manipulated by voice and gesture, was given at MIT in 1985.

2. Much as most of the buildings in the world go up without the contribution of an architect, now digital devices are often integrated into physical sites without any top-down design conception. Just as many builders regard the advice of an architect as one more opinion in a country where one is free to put things up however one pleases, many systems engineers regard the more human-centered advice of interaction designers as unscientific opinion in an area where technological expedience is enough.

3. On responsive spaces, two especially useful online resource pointer pages at the time of this writing were the Microsoft Research Intelligent Environments Resource Page (www.research.microsoft.com/ierp) and the National Institute of Standards and Technology Smart Spaces page (www.nist.gov/smartspace/ resources).

4. William Buxton, "Living in Augmented Reality: Ubiquitous Media and Reactive Environments," in K. Finn, A. Sellen & and S. Wilber eds., *Video-Mediated Communication.* Hillsdale, N.J.: Erlbaum, 1997, pp. 363–384.

5. Kumo Interactive, FX Palo Alto Laboratory. http://www.fxpal.com/smartspaces/. P. Chiu, A. Kapuskar, S. Reitmeier, and L. Wilcox, "Meeting Capture in a Media-Enriched Conference Room," in *Proceedings of the Second International Workshop on Cooperative Buildings (CoBuild '99). Lecture Notes in Computer Science,* Vol. 1670, New York: Springer-Verlag, 1999, pp. 79–88.

6. National Institute of Standards and Technology pointers on smart space research; http://www.nist.gov/smartspace/resources/.

7. This was a central theme of the e-boom. Besides the dematerialization of existing floors, new sites of exchanges arose around the "reverse market" principles of bidding. (eBay, FreeMarkets, Priceline, and Imandi were some sites at the time of this writing. Their prospects were uneven at best.)

8. Dan Siewiorek, et.al., "Adtranz: A Mobile Computing System for Maintenance and Collaboration," in *IEEE Wearable Computers 1998.* On virtual overlays for maintenance, see also http://www.automation.hut.fi/~etala/.

9. Steve Feiner, Webster, A., Krueger, T., MacIntyre, B., and Keller, E., "Architectural Anatomy," *Presence*, 4(3), pp. 318–325.

10. Karon MacLean, Scott Snibbe, and Golan Levin, "Tagged Handles: Merging Discrete and Continuous Manual Control." In Proceedings on

Computer-Human Interaction (CHI), The Hague: ACM, 2000. One application prototype addressed "deep parameter marking in graphic applications" (i.e., actions otherwise buried in menus). Another application used a similarly prepared wheel to advance through a movie, feeling bumps at the screen breaks.

11. Streetbox pollution monitor; Learian Design, http://www.citytech.co.uk/press_archive4.html.

12. For example, at the time this writing, BodyMedia.com was preparing to launch a service based on uploading bodily statistics to online analytical databases.

13. Mitchell, in *City of Bits*, accounted for the impact of connectivity on many of these, type by type.

14. A Carnegie Mellon design workshop project by Neil Werle and Margaret McCormack in 2000 explored this concept.

15. Home automation projects prominent at the time of this writing included National Outlook for Automation in the Home, at University of California, Irvine; Smart House, at MIT; the Aware Home, at Georgia Tech; the Intelligent House, at Lego; SmartHome.com, *Electronic House* (a magazine); Easy Living, at Microsoft; and the Adaptive House, at the University of Colorado, Boulder.

16. Ray Oldenburg, *The Great Good Place—Cafés, Coffee Shops, Community Centers, Beauty Parlors, General Stores, Bars, Hangouts, And How They Get You Through The Day*. New York: Paragon House, 1989.

17. As an illustration of how researchers will pursue whatever seems technically possible, consider smart beer glasses. In the iGlassware project at Mitsubishi Electronic Research Labs, Cambridge, each glass included a sensor and transmitter encased in dishwasher-safe coating that detected liquid level and identified the glass, plus a tabletop coil that gave power to the glass and sent signals to the bartender when the glass was empty. "Wireless Liquid Level Sensing for Restaurant Applications" www.merl.com/projects/iGlassware/.

18. A Carnegie Mellon design workshop project by Alex Nicolaidis and Eric Muehlmatt in 2000 explored this concept.

19. A Carnegie Mellon design workshop project by Andrea Klein and Laurence Yu in 2000 explored this concept.

20. A Carnegie Mellon design workshop project by Alex Nicolaidis and Iris Cho in 2000 explored this concept.

21. Chuihua Judy Chung, Jeffrey Inaba, Rem Koolhaas, and Sze Tsung Leong, eds., *The Harvard Design School Guide to Shopping*. Koln: Taschen, 2002.

22. In *Interpretations on Behalf of Place*, (Albany: State University of New York Press. 1994), Robert Mugeraurer described retail technology as a network of modular systems. Replaceable, disposable, "reserves in waiting," and channels of distribution are a global network in their own right.

23. Brain Opera. http://brainop.media.mit.edu. Also "The Story of the Brain Opera." http://lethe.media.mit.edu/text-site/libretto/story.html.

24. The SenseBus project, Interaccess, Toronto, 1999, http://www.interaccess.org/arg/sensebus/.

25. www.vindigo.com. This Palm navigation data service works by proximity. You enter an address, and it describes nearby features.

26. Keith Cheverst, Nigel Davies, Keith Mitchell, Adrian Friday and Christos Efstratiou. 2000. "Developing a Context-aware Electronic Tourist Guide: Some Issues and Experiences." In *Proceedings on Computer-Human Interaction (CHI).* The Hague, ACM, 2000. The Lancaster Guide Project, http://www.guide.lancs.ac.uk/overview.html

27. http://www.itsa.org/telematics.html "Because of their position on the windshield, these telematics mirrors, as they're often called, have an excellent view of the sky, and consequently are a great location for antennas, wireless modems and receivers. Automakers also like them because they can be installed across different vehicle platforms in a consistent location without redesigning instrument panels and overhead consoles, which already are crowded with features. Locating electronic features in the rearview mirror is also becoming more popular because the driver can view and interact with them while keeping his or her natural line of sight on the road ahead."

28. "Smart Intersection System to Ease Daily Traffic Flow," *Washington Post,* January 17, 2002.

29. The Electric Shoe Company, of Leicester, UK (www.theelectricshoeco.com) investigated the use of micro-electro-mechanical devices to generate small amounts of power, but enough to fuel a mobile phone, from the compression dynamics of footsteps. "These boots are made for talking," commented *Wired* magazine, June 28, 2000. www.wired.com/news/technology/0,1282,37276, 00.html)

Chapter 7

1. Herbert A. Simon, *The Sciences of the Artificial,* Cambridge: MIT Press, 1969, p. 82.

2. The interaction designer John Carroll has identified of six properties of conditions when design is necessary: (1) Incomplete description of the

problem to be addressed; (2) Lack of guidance on possible design *moves*; (3) The design *goal* or solution state cannot be known in advance; (4) *Trade-offs* among many interdependent elements; (5) Reliance on a *diversity* of knowledge and skills; (6) Wide ranging and ongoing impacts on human *activity*. (*Making Use–Scenario-Based Design of Human-Computer Interactions*, Cambridge: MIT Press, 2000, p. 22.)

3. The business role of design has been in play before. The "first machine age" of the early twentieth century saw the rise of design practices devoted to products rather than works, and design objects shaped by industrial process rather than cultural tradition. Much to the dismay of their critics, by the 1920s graphic and industrial design were perhaps also the first endeavors to elevate commercial interests to a level of formal consideration once reserved for patron-sponsored fine arts. Modern design came into its own in the service of corporate identity. Yearly model changes, widespread advertising campaigns, and of course the well-recognized logo became the recipients of the kinds of resources and talents that previous cultures devoted to mythic heros, religion, or the state.

4. Donald Schön, *The Reflective Practitioner—How Professionals Think in Action*. New York: Basic Books, 1983.

5. Explaining the limits of proceduralization, John Seeley Brown and Paul Duguid in *The Social Life of Information* (Cambridge: Harvard Business School Press, 2000) argued that "practice makes process," and not the other way around. What may seem like improvised workarounds can reveal contextual insight useful for incorporation in a process.

6. The question of subject matter may be critical to the independent notion of design. Richard Buchanan has explored this question as thoroughly as anyone. His essay "Wicked Problems in Design Thinking," which opens his edited volume *The Idea of Design,* (Cambridge: MIT Press, 1995) explains that while design is potentially universal in application, it does tend to arise more specifically when there is indeterminacy and planned artifice. Following from Horst Rittel's definition of "wicked problems", we can say that design problems lack explicitly definable limits. This is what makes subject matter so telling. What to take on is just as important as how to arrive at solutions.

7. David Kelley, "The Designer's Stance," in Terry Winograd, ed., *Bringing Design to Software*. Reading: Addison Wesley, 1996. IDEO's pioneering approach to information technology product design has been documented by Tom Kelley in *The Art of Innovation—Lessons in Creativity from IDEO, America's Leading Design Firm* (New York: Doubleday, 2001).

8. John Kao, *Jamming—The Art and Discipline of Business Creativity*. New York: Harper Business, 1996, p. 14.

9. Kao, *Jamming*, pp. 4, 13.

10. My own book, *Abstracting Craft* (Cambridge, MIT Press, 1996) was a response to this new perception.

11. In "Wicked Problems in Design Thinking," Buchanan classified the areas of design endeavor as symbolic and visual communication; material objects; activites and organized services; and complex systems of environments for living, working, playing, and learning.

12. Schön, *The Reflective Practicioner*.

13. Owing to this scientific bias for device efficiency without regard for human referents, interface design was a relative latecomer to the academy, and often emerged outside departments of computer science.

14. Alan Cooper, *The Inmates are Running the Asylum, or, Why High Technology Products Drive Us Crazy and How to Restore the Sanity*, Indianapolis: SAMS, 1999, p. 16. To a host of visionaries such as Cooper, interaction design is good business. For a representative set, see the many technology design thinkers represented Winograd, ed., *Bringing Design to Software*. At the beginning of the century, design thinking for computer scientists was advancing rapidly—but was still not being accepted in science-as-gospel circles based on precise, numerical methods.

15. The acronym GOMS was introduced by Stuart Card, Tom Moran, and Allan Newell in 1983, in their seminal book, *The Psychology of Human-Computer Interaction* (Hillsdale, N.J.: Erlbaum Associates.)

16. Jef Raskin, *The Humane Interface: New Directions For Designing Interactive Systems*. Reading: Addison Wesley, 2000, p. 5.

17. To the designer with an artistic education, this range raises the prospect of an avant-garde. This was perhaps expressed best in high-modernist critic Clement Greenberg's principle that to the masses at least, all great art is unappealing at first. Now some theorists extrapolate this, in what is heresy to usability orthodoxy, to claim that all great interfaces are unusable at first. Steve Cannon, formerly chief technical officer at Oven Digital, was one person to observe this.

18. Perhaps the most famous of these is lifestyle brands is Nike, which distinctly advertises sports, not sporting goods. Amid the 1996-1998 surge of the Nike swoosh, in which their logo appeared with a frequency and positioning normally reserved for traditional religious symbols, the brand appeared to be laying claim not only to particular sports, such as Tennis,, but to the active life itself.

19. Edward Tufte, *The Visual Display of Quantitative Information*. Cheshire, CT: Graphics Press, 1983.

20. Clement Mok, *Designing Business*. San Jose: Adobe Press, 1996, p. 46.

21. Clifford Geertz, *Interpretation of Cultures*. New York: Basic Books, 1973.

22. Hugh Beyer and Karen Holtzblatt. *Contextual Design—Defining Customer Centered Systems*. San Francisco: Morgan Kaufmann, 1998.

23. Beyer and Holtzblatt, *Contextual Design*, p. 47.

24. Beyer and Holtzblatt, *Contextual Design*, pp. 89–125.

25. Beyer and Holtzblatt, *Contextual Design*, pp. 89–125. This representation of work had five components: (1) Flow: an overview of communication between individuals and their responsibilities within an organization. (2) Sequence: different from flow in its emphasis on timing and its observation that tacit contextual conditions can trigger an explicit series of actions. (3) Artifacts: "a concrete trail of the work," which consists not only of objects of work but also the props and support structures. (4) Culture: standards, policies, and tacitly shared values that influence how things get done, and of which the tone of a setting is often the best indicator. (5) Physical environment: both constrains and reflects how an organization works.

26. Jennifer, Preece, Yvonne Rogers, and Helen Sharp, *Interaction Design—Beyond Human-Computer Interaction* (New York: John Wiley, 2002), a recent standard textbook for the field, devotes over half its chapters to observing, understanding, and evaluating users.

27. Preece, Sharp, and Rogers, *Interaction Design*. Design appropriateness depends on a host of factors beyond technical operation by the people who have their hands on the technology. Preece et al state this social context in terms of stakeholders. "Generally speaking, the group of stakeholders for a particular product is going to be larger than the group of people you'd normally think of as users.... Be very wary of changes which take power, influence, or control from some stakeholders without returning something in its place." (p. 171)

28. Preece et al. share my emphasis on expectations. "Involving users throughout development helps with expectation management because they have seen from an early stage what the product's capabilities are and what they are not." (Preece, Sharp, and Rogers, *Interaction Design*, p. 280.) I would add that they have seen what the design tradeoffs are, and what the opportunities for conceptual clarification might be; that is, why a particular design is advantageous in the first place.

29. Carroll, *Making Use*, has explained scenario planning in terms of experiences with software development.

30. For example, for some years the American Institute of Graphic Artists design summit conference titled itself "experience design." Web designer

Nathan Shedroff's book *Experience Design* (San Francisco: New Riders, 2001) was given a lot of exposure.

31. John Thackara, "The Design Challenge of Pervasive Computing." Keynote speech for ACM Computer-Human Interaction conference, 2000. www.doorsofperception.com.

32. William Mitchell, *E-Topia—Urban Life, Jim, but not as We Know It.* Cambridge: MIT Press, 1999.

33. Many of the finer texts on digital culture centered on questions of narrative. On narrative as a key to cultural prospects in digital media, one might put Janet Murray, Hamlet on the Holodeck, at one pole and Sven Birkerts, The Gutenberg Elegies at another. In between others such as Steven Johnson, Interface Culture, Mark Stefik, Internet Dreams, the Xerox PARC project and San Jose Museum exhbition, "experiments in the future of reading," describe the issues. Not only arts and letters, but also business planning and personal cultural life increasingly revolve around narrative. The more possibilities that exist, the more that people and organizations must shape some explanation of how they continue to aim, judge, or choose.

34. Carroll, *Making Use*, p. 26.

35. In his essay "On the Essential Contexts of Artifacts," in the volume *The Idea of Design*, Klaus Krippendorf generalized that "What something is (the totality of what it means) to someone consists of the sum total of its imaginable contexts. A knife has all kinds of uses; cutting is merely the most prominent one. Prying open a box, tightening a screw, scraping paint from a surface, cleaning dirty fingernails is as imaginable as picking a pickle from a jar. In the context of manufacturing, a knife is a cost. In the context of sales, a knife has exchange value. In the context of a hold-up, a knife may constitute a significant threat." (p. 159)

36. "We will come to think of interface design as a kind of art form, perhaps *the* art form, of the new century." (Steven Johnson, *Interface Culture—How New Technology Transforms the Way We Create and Communicate.* San Francisco: Harper Edge, 1997, p. 213.)

Chapter 8

1. Yi-Fu Tuan, "Place and Culture," in Wayne Franklin and Michael Steiner, eds., *Mapping American Culture*. Iowa City: University of Iowa Press, 1992, p. 29. "Place and culture can be viewed as a creative adaptation to the myriad individual human experiences of fragility and transience, aloneness and indifference. Place supports the human need to belong to a

meaningful and reasonably stable world, and it does so at different levels of consciousness, from an almost organic sense of identity that is an effect of habituation to a particular routine and locale, to a more conscious awareness of the values of middle-class places such as neighborhood, city, and landscape, to an intellectual appreciation of the planet earth itself as home." (p. 40)

2. About turf wars, academic discourse in cultural geography is a turf war in itself. Such politicized relativism has its limits. Phenomena exist that nobody can deny, such as sunrise, or the geological cuts and fills of the American interstate highway system. Some aspects of geography will be visible to the least constructed gaze long after the language games have ended. Thus even the least bookish person can take a drive on the outskirts of a town and not help reading a strange story in all that is being built so hastily out there.

3. Tuan, Yi-Fu, *Space and Place: The Perspective of Experience.* Minneapolis: University of Minnesota Press, 1976.

4. Edward Casey, *The Fate of Place.* Berkeley: University of California Press, 1997.

5. Casey, *The Fate of Place*, pp. 202–203. Among the many "ways into place," embodiment comes first. "Bodies and places are as inseparable as they are distinguishable." (p. 242)

6. Casey, *The Fate of Place*, p. 286.

7. Robert Mugerauer, in *Interpretations on Behalf of Place* (Albany: State University of New York Press, 1994, pp. 67–96), provided and de-politicized an overview of Heidegger's essential foundations on this topic.

8. Christian Norberg-Schulz, *Genius Loci—Towards a Phenomenology of Architecture,* New York: Rizzoli, 1983, p. 190.

9. Norberg-Schulz, *Genius Loci,* p. 190. "The development of individual and social identity is a slow process, which cannot take place in a continuously changing environment. We have every reason to believe the human alienation that is so common today, to a high extent is due to the scarce possibilities of orientation and identification offered by the modern environment." (p. 180).

10. The analogy between placelessness and bad weather was first brought to my attention by Rich Gold, at Xerox PARC.

11. J. B. Jackson, *Discovering the Vernacular Landscape.* New Haven: Yale University Press, 1984, p. 154.

12. Joshua Meyrowitz, *No Sense of Place: The Impact of Electronic Media on Social Behavior.* New York: Oxford University Press, 1985.

13. Meyrowitz, *No Sense of Place.*

14. M. Christine Boyer, "Disenchantment of the City," in *Cybercities*. Princeton: Princeton Architectural Press, 1995, p 119. Boyer's scathing attack on the marketing of Atlanta (or should that be Atlanta"?) cautions that whole cities could be subject to the kind of commodification, and its attendant trivialization and policing, that she has exposed in her many writings on South Street Seaport. "Imaging the City", in *Cybercities,* p. 148.

15. M. Christine Boyer, *The City of Collective Memory*. (Cambridge: MIT Press, 1994) is this critic's deeper work on the city as tableau and memory device, a topic that deserves more attention.

16. Anthony Vidler, *The Architectural Uncanny: Essays on the Modern Unhomely*. Cambridge: MIT Press, 1992, p. 163.

17. Vidler, *The Architectural Uncanny*, pp. 184–186. "The bond between 'body politic' and the city was, for the humanist tradition at least, more than a simple comparison. The psychological consequences of the loss of such a guiding metaphor can only be approached through the notion of homesickness, the desire to return to a once-safe interior." (p. 186)

18. Gupta, Akhil, and James Ferguson, eds. *Culture Power, Place— Explorations in Critical Anthropology*. Durham: Duke University Press, 1997. From their introduction: "It is fundamentally mistaken to conceptualize non- or supra-local identities (diasporic, refugee, migrant, national, etc.) as spatial and temporal extensions of a prior natural identity rooted in locality and community." (p. 7)

19. Edward Relph, *Place and Placelessness*. London: Pion, 1976, p. 141.

20. Relph's scale of insideness: (7) existential outsideness: alienation; (6) objective outsideness: deliberate detachment; (5) incidental outsideness: working, visiting; (4) vicarious insideness: through the arts; (3) behavioral insideness: as an outsider playing along; (2) empathetic insideness: openness, insight, and identification; (1) existential insideness: going native, being of the place. (*Place and Placelessness*, pp. 50–55.)

21. Relph, *Place and Placelessness,* pp. 82–83.

22. Relph, *Place and Placelessness,* p. 143.

23. These categories are not the wisdom of any particular anthropological luminary, but simply the consensus of several semesters of a seminar on place identity.

24. One example of an influential text text on globalization: Thomas Friedman, *The Lexus and the Olive Tree—Understanding Globalization*. New York: Farrar, Straus and Giroux, 2000.

25. www.iGo.com.

26. Jane Jacobs, *The Death and Life of Great American Cities*. New York: Random House, 1961, p. 222.

27. Bonnie Nardi and Vicki O'Day, *Information Ecologies—Using Technology with Heart*. Cambridge: MIT Press, 1999.

28. Nardi and O'Day, *Information Ecologies*.

29. The industrial ecologist Ezio Manzini suggested this illustrative Italian example of locality.

Chapter 9

1   Reduction to solely economic value is far from inconceivable. At the micro level, we all know places that discourage unplanned social interactions in favor of predictable consumer behavior. Regulations once posted at Universal Studios' City Walk in Los Angeles, for example, prohibited among other things, "non-commercial transactions."

2   Does a tract of forest have value by its very being? The USDA Forest Service motto for the national forests, "Land of many uses," reflects pure instrumentality. But today administrative attempts to reconcile recreational and extractive uses must play opposite the notion that a living place has value especially if no one uses it for anything at all.

3   Through theory, aesthetic value may also contrast with moral value. Centuries of provocateurs have made careers out of this difference.

4   John Ruskin, *Unto This Last: Four Essays on the First Principles of Political Economy*. London : A. C. Fifield, 1907.

5   Gerald Alonzo Smith. "The Purpose of Wealth: A Historical Perspective," in Herman Daly, ed., (1980), *Economics Ecology Ethics: essays toward a steady-state economy*.

6   John Ruskin, *Unto This Last*.

7   Smith, "The Purpose of Wealth."

8   It may seem obvious that you make money by helping other people make money, but designers need to learn and express this idea more often. Too often in the past, design was perceived as a way of helping other people *spend* money. In the sense that normative business practices understood design more in terms of its cost than its value, they enforced a cynical practice of it. Design that consisted of spending money trying to make preconceived entities more beautiful or usable after the fact was self-fulfilling in its unimportance.

9   Douglas Greenwald, ed., *Encyclopedia of Economics*. New York: McGraw-Hill, 1994.

10 John Dewey, "Theory of Valuation, part V: Ends and Values," in *Collected Works*, Vol. 13. Carbondale: Southern Illinois University Press, 1967, p. 220. (Original published in 1939.)

11 Dewey, "Theory of Valuation," p. 222. To some, this remark might appear as a complete denial of the possibility of faith.

12 Subjective utility goes beyond valuation by objective cost or price. It was on the question of utility that the classical nineteenth century theories of Smith, Ricardo, and Marx, which were based on cost of production, gave way to the neo-classical twentieth century economic theory based on market prices. A production-cost theory of value works well enough as long as traditional work is the normal mode of production, but it fails to explain value added by reorganization or capital investment. A software consultancy or an oil refinery is more profitable than a garment sweatshop, despite using far less labor.

13 Joseph Schumpeter, *A History of Economic Analysis*. New York: Oxford University Press, 1954.

14 A yard sale advertising that all items are reduced to one dollar will stop more cars than one advertising that everything is for free.

15 For a good orientation to the very existence of economics and its role at the start of the twenty-first century, from which the above has been paraphrased, see Kit Sims Taylor, "Human Society and the Global Economy," http://online.bcc.ctc.edu/econ100/ksttext/.

16 At least ever since the eighteenth century rationalist Jeremy Bentham proclaimed that a cardinal measure of utility could and should be established, economists have sought some sort of all-inclusive index of the greatest common good.

17 Not only can everyone can agree on dollars and cents, but also because people tend to mistake precision for accuracy, everyone lives with the fact that whatever is reported with greater numeric detail is taken more seriously. This is true even if vital factors are totally ignored by the measurements, calculations, and reports. The bottom line may be calculated out to the nearest hundred dollars on a million, but if it ignores human and cultural factors, it remains grossly incomplete.

18 This critique of supposedly value-free science was stated best by a scientist: Michael Polanyi, who in *Personal Knowledge* (Chicago: University of Chicago Press, 1958), made a lasting case for personal participation in the scientist's choice of what is measured.

19 Architect-scholar Michael Benedikt has approached the question of value from the perspective of the built environment, particularly as contrasted with dematerialization. He has reapproached value in terms of first principles of psychological economics and has come away with the assertion that value consists of an increase.

20 This expression is perhaps first attributable to the theologian and economist Matthew Fox in *The Reinvention of Work* (San Francisco: Harper
Perennial, 1994). The environmental economist Herman Daly, considered
by many to be a pioneer of that domain, once cast the problem in terms
of an ancient parable: "Humanity, craving the infinite, has been corrupt-
ed by the temptation to satisfy an insatiable hunger in the material realm.
Turn these stones into bread, urges Satan, and modern man sets to it ,
even to the extent of devising energy intensive schemes for grinding up
ordinary rock for materials—to eat the spaceship itself! But Jesus's
answer to the same temptation was more balanced: 'man does not live by
bread alone.' The proper object of economic activity is to have enough
bread, not infinite bread, not a world turned into bread, not even vast
storehouses of bread."

21 As Paul Hawken has pointed out, over three quarters of the world's economic production, as measured monetarily in gross domestic product,
involves only five percent of the world's people; and perhaps not coinci-
dentally, nearly one person in six worldwide is unemployed or underem-
ployed. (*The Ecology of Commerce: A Declaration of Sustainability*, New
York: Harper Collins, 1993.)

22 For better or worse in this new century, life itself becomes the goods.
Some swing toward intrinsic value seems inevitable in this regard.
Valuation of the *causa materialis*, the life-sustaining substrates ignored by
pragamatic industrialism because there was no need, no change, no mar-
ginal utility, returns to the limelight. The perennial pursuit of a universal
measure takes new forms, in which design has a role. For example,
Michael Benedikt, whose constructs for cyberspace led him into a thorough reconsideration of value, has positioned design as complexity and
organization. "Life-time," the product of life and time, which is very
much like Aristotle's "human flourishing" with some room made for non-
humans. Life-time is a sort of vitalist, Bergsonian idea that what really,
ultimately, is of most value is that which promotes life—in the sense of
alive-ness—at the highest levels of complexity and organization, over the
largest number of living things, and for the longest time."
Michael Benedikt. Value: Psychoeconomics/Value/Money/Architecture/...
Excerpts from a talk given at the first meeting of the symposium series On
The Question of Economic Value. 1994. Appeared in *platFORM
Architecture News*(School of Architecture, University of Texas at Austin),
Fall 1994.

23 Paul Hawken, *The Ecology of Commerce—A Declaration of Sustainability.*
New York: Harper Collins, 1993.

24 Manuel Castells, *The Internet Galaxy—Reflections on the Internet, Business, and Society*. New York, Oxford University Press, 2001.

25 Jon Logan and Harvey Molotch, *Urban Fortunes—The Political Economy of Place*. Berkeley: University of California Press, 1987, pp. 43, 9, 18, 9, 18, 262.

26 In the near-futurist cyberpunk novel *Snow Crash* (New York: Bantam, 1992), the novelist Neal Stephenson characterized America's ultimate specialties as "music, movies, and microcode (also high speed pizza delivery)." (p. 2)

27 Sadly, consumerism's worst victims are exactly the ones who require more or less constant entertainment. Was music, humanity's greatest expression, really meant to come out of gas pumps and restroom ceilings?

Chapter 10

1 Janine Benyus, (1997), *Biomimicry: Innovation Inspired by Nature*.

2 Gary Snyder, "The Place the Region and the Commons" in *The Practice of the Wild*. San Francisco : North Point Press, 1990, pp. 25-47. Quote on p. 39.

3 Snyder, "The Place the Region and the Commons," p. 40.

Further Reading

Here follow a dozen suggestions of books that were essential resources to this one, and which may interest general readers.

Brand, Stewart. 1994. *How Buildings Learn—What Happens after they are Built.* New York: Penguin. On types and patterns, and arguably the best book on architecture to come out of the 1990s.

Dourish, Paul 2001. *Where the Action Is.* Cambridge: MIT Press. Deep philosophical foundations for embodiment in interaction design.

Golledge, Reginald, and Robert Stimson. 1997. *Spatial Behavior—A Geographic Perspective.* New York: Guilford Press. A full compendium of the analytical academic discipline of environment-and-behavior.

Hawken, Paul, Amory Lovins and Hunter Lovins. 1999. *Natural Capitalism—Creating the Next Industrial Revolution.* Boston: Little, Brown. The most visible explanation of industrial ecology.

Laurel, Brenda. 1991. *Computers as Theater.* Reading, MA: Addison Wesley. A foundational book in interaction design.

Mitchell, William J. 1999. *e-Topia—Urban Life, Jim, but not as We Know It.* Cambridge: MIT Press. The more sober companion to the earlier *City of Bits*, with relevant emphasis on "computers for living in," and a full-scale agenda of "lean and green."

Moran, Tom and Paul Dourish, eds. 2001. *Special issue: Context-Aware Computing, V.16, n. 2–4, of Human Computer Interaction.* Mahweh, NJ: Lawrence Erlbaum Associates. A solid portrait of the state of the field at the time, including but not limited to technical matters.

Preece, Jennifer, Yvonne Rogers, Helen Sharp. 2002. *Interaction Design—Beyond Human-Computer Interaction.* New York: John Wiley. An instant standard textbook in its field, with particular emphasis on the necessity but insufficiency of usability studies.

Rossi, Aldo. 1982 (1964). Trans. Diane Ghirardo. *The Architecture of the City.* Cambridge: MIT Press. The seminal declaration of typology in architecture.

Shafer, Steve, "Ten Dimensions of Ubiquitous Computing." In Nixon, Paddy, Gerard Lacey, and Simon Dobson, eds. 2000. *Managing Interactions in Smart Environments (MANSE '99)*. London: Springer-Verlag. www.research.microsoft.com/projects/easyliving/. The best overview of the technology at the time, and accompanied by various significant technical papers.

Thackara, John. 2005. *In the Bubble.* An insightful collection of essays on pervasive computing, industrial ecology, and design culture. Accumulated insight from the legendary design conferences, Doors of Perception.

Tuan, Yi-Fu. 1976. *Space and Place: The Perspective of Experience.* Minneapolis: University of Minnesota Press. The most concise summary of Tuan's unequalled interpretation of environmental psychology.

References

Agre, Philip. 2001. "Changing Places: Contexts of Awareness in Computing." In Tom Moran and Paul Dourish, eds. Special issue: "Context-Aware Computing," V.16, n. 2–4, of *Human Computer Interaction*. Mahweh, N.J.: Erlbaum Associates.

Alexander, Christopher. 1977. *A Timeless Way of Building*. New York: Oxford University Press.

Alexander, Christopher, Sara Ishikawa, and Murray Silverstein, with Max Jacobson, Ingrid Fiksdahl-King, and Shlomo Angel. 1977. *A Pattern Language—Towns, Buildings, Construction*. New York: Oxford University Press.

Augé, Mark. 1995. *Non-places: Introduction to the Anthropology of Supermodernity*. London: Verso.

Aukstakalnis, Steve, David Blattner and Steve Roth. 1992. *Silicon Mirage—The Art and Science of Virtual Reality*. Berkeley: Peachpit Press.

Biegl, Michael, Phil Gray, and Daniel Slaber, eds. 2001. *Proceedings of the Workshop on Location Modeling—Ubicomp 2001*.

Benedikt, Michael (ed), 1992. *Cyberspace: First Steps*. Cambridge: MIT Press.

Benjamin, Walter. 1935. "The Work of Art in the Age of Mechanical Reproduction," in Hannah Arendt, ed., 1969, *Illuminations*. New York: Schochen Books.

Benyus, Janine. 1997. *Biomimicry: Innovation Inspired by Nature*. New York: Morrow.

Berry, Wendell. 1977. *The Unsettling of America*. San Francisco: Sierra Club Books.

Beyer, Hugh and Karen Holtzblatt. 1998. *Contextual Design—Defining Customer Centered Systems*. San Francisco: Morgan Kaufmann.

Bloomer, Kent, and Charles Moore. 1977. *Body, Memory, and Architecture*. New Haven: Yale University Press.

Borchers, Jan. 2001. *A Pattern Approach to Interaction Design*. New York: John Wiley.

Boriello, Gaetano and Roy Want. 2000. "Embedded Computation Meets the World Wide Web." *Communications of the ACM*, 43(May 2000): 5.

Boyer, Christine. 1994. *The City of Collective Memory*. Cambridge: MIT Press.

Boyer, Christine. 1995. *CyberCities*. Princeton: Princeton Architectural Press.

Brand, Stewart. 1994. *How Buildings Learn—What Happens after they are Built*. New York: Penguin.

Brumitt, Barry, John Krumm, Brian Meyers, Steven Shafer. 1999. "Easy Living: Ubiquitous Computing & The Role of Geometry." Microsoft: research. microsoft.com/easyliving

Brumitt, Barry. 2002. "Microsoft Research: Intelligent Environments Resource Page." www.research.microsoft.com/ierp/

Brown, John Seely, and Paul Duguid. 2000. *The Social Life of Information*. Cambridge: Harvard Business School Press.

Brown, John Seely, and Mark Weiser. 1996. "The Coming Age of Calm Technology." www.ubiq.com/hypertext/weiser/acmfuture2endnote.htm

Buchanan, Richard, and Victor Margolin, eds. 1995. *Discovering Design*. Chicago: University of Chicago Press.

Buchanan, Richard, ed., 1995. *The Idea of Design*. Cambridge: MIT Press.

Buxton, William. 1997. "Living in Augmented Reality: Ubiquitous Media and Reactive Environments." In K. Finn, A. Sellen and S. Wilber, eds. *Video Mediated Communication*. Hillsdale, N.J.: Erlbaum, 363–384.

Buxton, William. 1995. "Integrating the Periphery and Context: A New Taxonomy of Telematics." http://www.dgp.toronto.edu/people/rroom/research/papers/ bg_fg/bg_fg.html.

Campbell, Joseph. 1988. *The Power of Myth*. New York: Doubleday.

Card, Stuart, Tom Moran, and Alan Newell. 1983. *The Psychology of Human-Computer Interaction*. Hillsdale, N.J.: Erlbaum Associates.

Carroll, John. 2000. *Making Use—Scenario-Based Design of Human-Computer Interactions*. Cambridge: MIT Press.

Casey, Edward. 1997. *The Fate of Place*. Berkeley: University of California Press.

Castells, Manuel. 1989. *The Informational City*. Oxford: Blackwell.

Castells, Manuel. 2001. *The Internet Galaxy—Reflections on the Internet, Business, and Society*. New York: Oxford University Press.

Cheverst, Keith, Nigel Davies, Keith Mitchell, Adrian Friday and Christos Efstratiou. 2000. "Developing a Context-aware Electronic Tourist Guide: Some Issues and Experiences." *Proceedings on Computer-Human Interaction (CHI)*. The Hague. ACM.

Chung, Chuihua Judy, Jeff Inaba, Rem Koolhaas, Sze Tsung Leong, eds., (2002), *The Harvard Design School Guide to Shopping*. London: Taschen.

Cooper, Alan. 1999. *The Inmates are Running the Asylum, or, Why High Technology Products Drive Us Crazy and How to Restore the Sanity*. Indianapolis: SAMS.

Cooper, Alan. 1995. *About Face: the Essentials of Interface Design*. Foster City, CA.: IDG.

Daly, Herman, ed. 1980. *Economics Ecology Ethics: Essays toward a Steady-State Economy*. San Francisco : W.{t}H. Freeman.

Dewey, John. 1967 (1939). "Theory of Valuation." V: "Ends and Values." *Collected Works*. Carbondale: Southern Illinois University Press.

Doheny-Farina, Stephen. 1996. *The Wired Neighborhood*. New Haven: Yale Press.

Douglas, Mary. 1986. *How Institutions Think*. Syracuse: Syracuse University Press.

Dourish, Paul 2001. *Where the Action Is*. Cambridge: MIT Press.

Downs, Robert and David Stea. 1973. *Image and the Environment: Cognitive Mapping and Spatial Behavior*. Chicago: Aldine Press.

Dreyfuss, Henry. 1955. *Designing for People*. New York, Simon and Schuster.

Dreyfus, Hubert. "The Current Relevance of Merleau-Ponty's Phenomenology of Embodiment." In Honi Haber and Gail Weiss (eds.) (1996). *Perspectives on Embodiment*, New York: Routledge. pp. 103–120.

Duffy, Francis. 1997. *The New Office*. London: Conran.

Dunne, Anthony and Fiona Raby. 2001. *Design Noir—The Secret life of Electronic Objects*. Princeton: Princeton Architectural Press.

Eilan, Naomi, Rosaleen McCarthy, and William Brewer (eds). 1993. *Spatial Representation—Problems in Philosophy and Psychology*. Oxford: Blackwell.

Eliade, Mircea. 1959. *The Sacred and the Profane*. New York: Harcourt Brace.

Enge, Per. 2002. "Locator madness pervades plenty of devices." CNN Sci-tech, March 28.

Ernst, Jennifer, et. al., (eds.) 1995. *The PARC Story*. Palo Alto: Xerox.

Estrin, Deborah, Ramesh Govindan, and John Heidemanm, eds. Introduction to "Embedding the Internet." Issue focus of *Communications of the ACM*. 43(May 2000): 5.

Evenson, Norma. 1979. *Paris—A Century of Change, 1878–1978*. New Haven: Yale Press.

Feiner, S., Webster, A., Krueger, T., MacIntyre, B., and Keller, E. "Architectural Anatomy." *Presence* 4(1995): 318–325. Also ww.cs.columbia.edu/graphics/ projects/archAnatomy/architecturalAnatomy.html.

Foster, Hal. 1996. *The Return of the Real*. Cambridge: MIT Press.

Fox, Matthew. 1994. *The Reinvention of Work*. San Francisco: Harper Perennial.

Michael Fox and Bryant Yeh. 2000. "Intelligent Kinetic Systems in Architecture," in Nixon, Lacey, and Dobson, eds., *Managing Interactions in Smart Environments (MANSE '99)*. London: Springer-Verlag.

Franklin, Wayne, and Michael Steiner, eds. 1992. *Mapping American Culture*. Iowa City: University of Iowa Press.

Friedman, Diana, ed. 1996 "Community—Just what is all the talk about?" *Metropolis*, October 1996.

Thomas Friedman. 2000. *The Lexus and the Olive Tree—Understanding Globalization*. New York: Farrar, Straus and Giroux.

Gallagher, Winifred. 1993. *The Power of Place—How Our Surroundings Shape Our Thoughts, Feelings, and Actions*. New York: Poseidon Press.

Garfinkel, Simson. 2000. *Database Nation: The Death of Privacy in the 21st Century*. Cambridge: O'Reilly.

Gassmann, Oliver, and Hans Meixner, eds. 2001. *Sensors in Intelligent Buildings*. New York: John Wiley.

Gelertner David. 1995. *1939, The Lost World of the Fair*. New York: Free Press.

Gibson, James J. 1979. *The Ecological Approach to Visual Perception*. Boston: Houghton Mifflin.

Geertz, Clifford. 1973. *Interpretation of Cultures*. New York, Basic Books.

Gibson, William. 1984. *Neuromancer*. New York: Ace Books.

Giedion, Sigfried. 1941. *Space, Time, and Architecture—The Growth of a New Tradition*. Cambridge: Harvard University Press.

Golledge, Reginald, and Robert Stimson. 1997. *Spatial Behavior—A Geographic Perspective*. New York: Guilford Press.

Graham, Steven, and Simon Martin. 2001. *Splintering Urbanism—Networked Infrastructures, Technological Mobilities, and the Urban Condition*. London: Routledge.

Graham, Steven, and Simon Martin. 1996. *Telecommunications and the City—Electronic Spaces, Urban Places*. London: Routledge.

Gregory, Richard, and O.{t}L. Zangwill, eds. 1998. *The Oxford Companion to the Mind*. New York: Oxford University Press.

Greenwald, Douglas, ed. 1994. *Encyclopedia of Economics*. New York: McGraw-Hill.

Guimbretière, François, Maureen Stone, and Terry Winograd. 2001. "Fluid Interaction with High-Resolution Wall-size Displays." *Proceedings of UIST 2001*. New York: ACM Press, pp. 21–30.

Gupta, Akhil, and James Ferguson, eds. 1997. *Culture Power, Place—Explorations in Critical Anthropology*. Durham: Duke University Press.

Hall, Edward. 1966. *The Hidden Dimension*. New York: Doubleday.

Hardin, Garrett. 1968. "The Tragedy of the Commons." *Science* 162(1968): 1243–1248.

Harrison, Helen, ed. 1980. *Dawn of a New Day—The New York World's Fair, 1939/40*. Catalogue of the Queens Museum exhibit. New York: New York University Press.

Harrison, Steve, and Paul Dourish. 1996. "Re-Placing Space: The roles of Place and Space in Collaborative Systems." *Proceedings of CSCW '96*.

Hawken, Paul. 1993. *The Ecology of Commerce: A Declaration of Sustainability*. New York: Harper Collins.

Hawken, Paul, Amory Lovins and Hunter Lovins. 1999. *Natural Capitalism—Creating the Next Industrial Revolution*. Boston: Little, Brown.

Hawking, Steven and Roger Penrose. 2000. *The Nature of Space and Time*. Princeton: Princeton University Press.

Heschong, Lisa. 1979. *Thermal Delight in Architecture*. Cambridge: MIT Press.

Hughes, Robert. 1992. "Art, Morals, and Politics." *New York Review of Books*. April 23, 1992.

Ingold, Tim. 1986. *The Appropriation of Nature. Essays on Human Ecology and Social Relations*. Manchester, UK: Manchester University Press.

Ishii, Hiroshi, and Brygg Ullmer. 1997."Tangible Bits: Towards Seamless Interfaces between People, Bits and Atoms," *Proceedings of Conference on Human Factors in Computing Systems (CHI '97)*, Atlanta: ACM, pp. 234–241.

Ishii, Hiroshi, C. Wisnecki, S. Brave, A. Dahley, M. Gorbet, B. Ullmer, and P. Yarin. 1998. "ambientROOM: Integrating Ambient Media with Architectural Space," *Proceedings of Conference on Human Factors in Computing Systems (CHI '98)*, New York: ACM, pp. 173–174.

Jackson, J.{t}B. 1984. *Discovering the Vernacular Landscape.* New Haven: Yale University Press.

Jacobs, Jane. 1961. *The Death and Life of Great American Cities.* New York: Random House.

Johnson, Steven. 1997. *Interface Culture—How New Technology Transforms the Way We Create and Communicate.* San Francisco: Harper Edge.

Kao, John. 1996. *Jamming: The Art and Discipline of Business Creativity.* New York: Harper Business.

Kelbaugh, Douglas. 2002. *Repairing the American Metropolis—Common Place Revisited.* Seattle: University of Washington Press.

Kelley, Tom. 2001. *The Art of Innovation—Lessons in Creativity from IDEO, America's Leading Design Firm.* New York: Doubleday.

Kunstler, James. 1993. *The Geography of Nowhere.* New York: Touchstone.

Lakoff, George, and Mark Johnson. 1980. *Metaphors We Live By.* Chicago: University of Chicago Press.

Lakoff, George, and Mark Johnson. 1999. *Philosophy in the Flesh—The Embodied Mind and Its Challenge to Western Thought.* New York: Basic Books.

Laurel, Brenda. 1989. *The Art of Human-Computer Interface Design.* Reading, Mass.: Addison Wesley.

Le Corbusier, and Francois Pierrefeu. 1942. *Maison des Hommes.* Paris: Plon.

Lefebvre, Henri. 1991 (1974). Trans. Donald Nicholson-Smith. *The Production of Space.* Oxford: Basil Blackwell.

Levy, David. 1994. "Fixed or Fluid? Document Stability and New Media." In *European Conference on Hypertext Technology '94 Proceedings.* Edinburgh: ACM.

Logan, Jon and Harvey Molotch. 1987. *Urban Fortunes—The Political Economy of Place.* Berkeley: University of California Press.

Lynch, Kevin. 1960. *The Image of the City.* Cambridge: MIT Press.

Machover, Tod, et. al. 1995. "Brain Opera." http://brainop.media.mit.edu/

MacLean, Karon, Scott Snibbe, and Golan Levin. 2000. "Tagged Handles: Merging Discrete and Continuous Manual Control." In Proceedings on Computer-Human Interaction (CHI) The Hague. ACM.

Marcus, Clare Cooper. 1995. *House as Mirror of Self.* Berkeley: Conari Press.

Marx, Leo. 1964. *The Machine in the Garden—Technology and the Pastoral Ideal in America.* New York: Oxford University Press.

McCarthy, Joseph, and Eric Meidel. 1999. "ActiveMap: A Visualization Tool for Location Awareness to Support Informal Interactions." In Gellerson, ed., *Handheld and Ubiquitous Computing.* Berlin: Springer-Verlag.

McCullough, Malcolm. 1996. *Abstracting Craft.* Cambridge: MIT Press.

McHarg, Ian. 1969. *Design with Nature.* New York: Garden City.

McPhee, John. 1989. *The Control of Nature.* New York: Farrar, Straus, Giroux.

Merleau-Ponty, Maurice. 1945. Colin Smith, trans., 1962. *Phenomenology of Perception.* London: Routledge.

Meyrowitz, Joshua. 1985. *No Sense of Place: The Impact of Electronic Media on Social Behavior.* New York: Oxford University Press.

Mitchell, William J. 1995. *City of Bits—Space, Place, and the Infobahn.* Cambridge: MIT Press.

Mitchell, William J. 1999. *e-Topia—Urban Life, Jim, but not as We Know It.* Cambridge: MIT Press.

Mok, Clement. 1996. *Designing Business.* San Jose: Adobe Press.

Moran, Tom and Paul Dourish, eds. 2001. *Special issue: Context-Aware Computing, V.16, n. 2–4, of Human Computer Interaction.* Mahweh, N.J.: Erlbaum Associates.

Moore, Gary and Reginald Golledge, eds, 1976. *Environmental Knowing— Theories, Research, Methods.* Stroudsburg, Pa.: Dowden, Hutchison, Ross.

Mugerauer, Robert. 1994. *Interpretations on Behalf of Place.* Albany: State University of New York Press.

Mumford, Lewis. 1936. *Technics and Civilization.* New York: Harcourt, Brace.

Nardi Bonnie, ed., 1996. *Context and Consciousness: Activity Theory and Human-Computer Interaction.* Cambridge: MIT Press.

Nardi, Bonnie and Vicki O'Day. 1999. *Information Ecologies—Using Technology with Heart.* Cambridge: MIT Press.

Nixon, Paddy, Gerard Lacey, and Simon Dobson, eds. 2000. *Managing Interactions in Smart Environments (MANSE '99).* London: Springer-Verlag.

Norberg-Schulz, Christian. 1983. *Genius Loci—Towards a Phenomenology of Architecture.* New York: Rizzoli.

Norman, Don. 1998. *The Invisible Computer.* Cambridge: MIT Press.

Nye, David. 1992. *Electrifying America—Social Meanings of a New Technology, 1880–1940.* Cambridge: MIT Press.

Nye, David. 1994. *American Technological Sublime.* Cambridge: MIT Press.

Oldenburg, Ray. 1989. *The Great Good Place: Cafés, Coffee Shops, Community Centers, Beauty Parlors, General Stores, Bars, Hangouts, And How They Get You Through The Day.* New York: Paragon House.

Pentland, Alex. 2000. "Perceptual Intelligence." *Communications of the ACM* 43(March, 2000): 1.

Polanyi, Michael. 1958. *Personal Knowledge.* Chicago: University of Chicago Press.

Pool, Ithiel de Sola. 1977. *The Social Impact of the Telephone.* Cambridge: MIT Press.

Preece, Jennifer, Yvonne Rogers, Helen Sharp. 2002. *Interaction Design—Beyond Human-Computer Interaction.* New York: John Wiley.

Price, Jennifer. 1999. *Flight Maps—Adventures with Nature in Modern America.* New York: Basic Books.

Putnam, Hilary. 1988. *Representation and Reality.* Cambridge: MIT Press.

Raskin, Jef. 2000. *The Humane Interface: New Directions For Designing Interactive Systems.* Reading, Mass.: Addison Wesley.

Reeves, Byron and Clifford Naas. 1996. *The Media Equation—How People Treat Computers, Television, and New Media Like Real People and Places.* Cambridge, UK: Cambridge University Press.

Rekimoto, Jun and Masanori Saitoh. 1999. "Augmented Surfaces: A Spatially Continuous Work Space for Hybrid Computing Environments." *Proceedings of Conference on Human Factors in Computing Systems (CHI '99)*

Relph, Edward. C. 1976. *Place and Placelessness.* London: Pion.

Rheingold, Howard. 1994. *Virtual Communities—Homesteading on the Electronic Frontier.* New York: Harper Perennial.

Rose, H.{t}J. 1959. *A Handbook of Greek Mythology.* New York: E.{t}P. Dutton.

Rossi, Aldo. 1982 (1964). *The Architecture of the City.* Cambridge: MIT Press.

Ruskin, John. 1907. *Unto This Last: Four Essays on the First Principles of Political Economy.* London: A.{t}C. Fifield.

Schön, Donald. 1983. *The Reflective Practitioner—How Professionals Think in Action.* New York: Basic Books.

Schama, Simon. 1987. *The Embarrassment of Riches*—An Interpretation of Dutch Culture in the Golden Age. New York: Knopf.

Schumpeter, Joseph. 1954. *A History of Economic Analysis*. New York: Oxford University Press.

Searle. John, 1992. *The Rediscovery of the Mind*. Cambridge: MIT Press.

Shafer, Steve. 2000. "Ten Dimensions of Ubiquitous Computing." In Nixon, Paddy, Gerard Lacey, and Simon Dobson, eds., *Managing Interactions in Smart Environments (MANSE '99)*. London: Springer-Verlag. www.research. microsoft.com/projects/easyliving/.

Shedroff, Nathan. 2001. *Experience Design*. San Francisco: New Riders. www.nathan.com/thoughts/.

Shenk, David. 1997. *Data Smog: Surviving the Information Glut*. San Francisco: Harper Edge.

Shepheard, Paul. 1997. *The Cultivated Wilderness, or, What is Landscape?* Cambridge: MIT Press.

Shilling, Chris. 1993. *The Body and Social Theory*. London: Sage.

Siegel, A.{t}W. and S.{t}H. White. 1975. "The Development of Spatial Representations of Large-Scale Environments," in *Advances in Child Development and Behavior*. New York: Academic Press, Vol. 10, pp. 10–55.

Siewiorek, Dan, et.al., "Adtranz: A Mobile Computing System for Maintenance and Collaboration," in *IEEE Wearable Computers 1998*.

Simon, Herbert A. 1969. *The Sciences of the Artificial*. Cambridge: MIT Press.

Snyder, Gary. 1990. *The Practice of the Wild*. San Francisco: North Point Press.

Spretnak, Charlene. 1997. *The Resurgence of the Real—Body, Nature and Place in a Hypermodern World*. New York: Routledge.

Stephenson, Neal. 1992. *Snow Crash*. New York: Bantam.

Stilgoe, John. 1983. *Metropolitan Corridor—Railroads and the American Scene*. New Haven: Yale Press.

Streitz, Norbert, Jane Siegel, Volker Hartkopf, and Shin'ichi Konomi, eds. *Cooperative Buildings—Integrating Information, Organization, and Architecture. Second International Workshop: Co-Build '99*. Berlin: Springer-Verlag.

Suchman, Lucy. 1986. *Plans and Situated Actions*. Cambridge, UK: Cambridge University Press.

Taylor, Kit Sims. 2001. "Human Society and the Global Economy." http://online.bcc.ctc.edu/econ100/ksttext/.

Tennenhouse, David. 2000. "Proactive Computing." *Communications of the ACM*. 43:5 May 2000.

Tenner, Edward. 2000. *Why Things Bite Back—Technology and the Revenge of Unintended Consequences*. New York: Vintage.

Thackara, John. 2002. "The Design Challenge of Pervasive Computing." Introduction to the conference, Doors of Perception 7: *Flow*. www.doorsofperception.com.

Thackara, John. 1995– . In the Bubble. Collected essays at www.doorsofperception.com/In+the+Bubble/.

Tristam, Claire. 2001. "The Next Computer Interface," *Technology Review*, December 2001.

Tuan, Yi-Fu. 1979. *Landscapes of Fear*. Minneapolis: University of Minnesota Press.

Tuan, Yi-Fu. 1976. *Space and Place: The Perspective of Experience*. Minneapolis: University of Minnesota Press.

Tufte, Edward. 1983. *The Visual Display of Quantitative Information*. Cheshire, Conn.: Graphics Press.

Tversky, Barbara. 1991. "Spatial Mental Models." In *The Psychology of Learning and Motivation*. New York: Academic Press, Vol. 27, pp. 109–145.

Tversky, Barbara. 1992. "Memory for Pictures, Environment, Maps, and Graphs." In Payne and Conrad, eds., *Practial Aspects of Memory*. Hillsdale, N.J.: Lawrence Erlbaum.

Vidler, Anthony. 1992. *The Architectural Uncanny: Essays in the Modern Unhomely*. Cambridge: MIT Press.

Walker, John 1989. "Through the Looking Glass," in *The Art of Human Computer Interaction*. Reading, Mass.: Addison Wesley.

Weibel, Peter, and Christof Schmid, eds. 2001. CTRL[SPACE]. Catalogue of the International media\art award 1991. Karlsruhe: ZKM.

Weiser, Mark. 1991. "Future Computers." In *Scientific American* 265:3. Special Issue: *The Computer in the 21st Century*, September 1991.

Weiser,. Mark, and John Seeley Brown and 1996. "The Coming Age of Calm." www.ubiq.com/hypertext/weiser/acmfuture2endnote.htm.

Weiss, Michael. 1988. *The Clustering of America*. New York: Harper and Row.

Weizenbaum, Joseph. 1976. *Computer Power and Human Reason*. San Francisco: Freeman.

Wellner, Pierre, Wendy Mackay, Rich Gold. 1993. "(Back to the) Real World." *Communications of the ACM* 36(July 1993): 7.

White, E.{t}B. 1939. "The World of Tomorrow." Originally appeared in *The New Yorker;* now in print in the 1977 collection Essays of E.{t}B. White, New York: Harper and Row, pp. 139–148.

Wiener, Norbert. 1948. *Cybernetics; or, Control and Communication in the Animal and the Machine*. Cambridge: Technology Press.

Winograd, Terry, ed. 1996. *Bringing Design to Software*. Reading: Addison Wesley.

Wood, Denis. 1992. *The Power of Maps*. New York: Guilford Press.

Wooley, Benjamin. 1993. *Virtual Worlds*. London: Penguin.

Yarin, Paul, and Hiroshii Ishii. 1999. "TouchCounters: Desiging Interactive Electronic Labels for Physical Containers." *In Proceedings on Computer-Human Interaction (CHI)*. New York: ACM.

Zuboff, Shoshana. 1988. *In the Age of the Smart Machine—The Future of Work and Power*. New York: Basic Books.

Index

"This is an important book. Not so much for what it achieves for architecture specifically, but for its detailed scholarly critique of the present level of ubiquitous, embedded computing devices generally."
—David Harle, *Leonardo Digital Reviews*

"Malcolm McCullough's new book *Digital Ground: Architecture, Pervasive Computing, and Environmental Knowing* is a readable and timely contribution to current interaction design. Using ideas drawn from architectural and design theory, cognitive science, and philosophy, McCullough significantly extends current ideas about pervasive computing and so-called experience design, while building on the foundation of traditional task-centered interface design. It's the best current book on interaction design, and should appeal to both designers and theorists . . . *Digital Ground* should be required reading for anyone interested in where interaction design is headed."
—Andrew Otwell *heyotwell*

"I bought and read one standout book this year, Malcolm McCullough's *Digital Ground*, mixed in with many more that I enjoyed. *Digital Ground* stood out as it combined a lot of things I had been thinking about, but had not quite pulled together."
—Thomas Vander Wal, *vanderwal.net, full review*

"In *Digital Ground* Malcolm McCullough elegantly summarizes the past and present relations between architecture and computing, and constructs a solid foundation for future interaction between the two fields."
—Casey Reas, Interaction Design Institute Ivrea

"This is one of the most thoughtful books in the emerging field of interaction design. It is well argued and solidly grounded in both the literature and experience of computing. McCullough provides a powerful explanation for why design — and interaction design in particular — is emerging as a liberal art of the twenty-first century. *Digital Ground* is important for the professional designer, the student of design, and the general public."
—Richard Buchanan, Carnegie Mellon University

"Malcolm McCullough's book charts a significant, unexplored terrain confronting architects and society at large. Pervasive computing is embedded, networked, ubiquitous, and capable of not only sensing and processing, but acting as well. This new form of computing holds the potential to restructure physical space and our relation to it, and McCullough provides an articulate and readable introduction to this new world, both promising and troubling. *Digital Ground* is a solid, early contribution to what will quickly become an important field of study for architecture, planning, and urban design."
—Dana Cuff, Professor of Architecture, University of California, Los Angeles